ALLIS-CHALMERS

Original
ALLIS-CHALMERS
1933–1957

Guy Fay and Andy Kraushaar

MBI Publishing Company

First published in 2000 by MBI Publishing Company, 729 Prospect Avenue, PO Box 1, Osceola, WI 54020-0001 USA

MBI Publishing Company books are also available at discounts in bulk quantity for industrial or sales-promotional use. For details write to Special Sales Manager at Motorbooks International Wholesalers & Distributors, 729 Prospect Avenue, PO Box 1, Osceola, WI 54020-0001 USA.

Library of Congress Cataloging-in-Publication Data

Fay, Guy.
Original Allis-Chalmers, 1933–1957 / Guy Fay & Andy Kraushaar.
p. cm.
Includes index.
ISBN 0-7603-0439-4 (hc. : alk. paper)
1. Allis-Chalmers tractors—History.
I. Kraushaar, Andy. II. Title.
TL233.6.A43 F39 2000
629.225'2—dc21 00-058407

On the front cover: Paul Demler's 1944 styled WC. *Andy Kraushaar*

On the frontispiece: Detail of Larry and Edwin Karg's 1950 Model C, which features an aftermarket gearbox that Allis-Chalmers experimented with. *Andy Kraushaar*

On the title page: Larry and Edwin Karg's 1950 Model C. *Andy Kraushaar*

On the back cover: Photos by Andy Kraushaar

Edited by Steve Hendrickson
Designed by Laura Henrichsen

Printed in China

Contents

ALLIS-CHALMERS

Preface

The author grew up on a farm with Allis-Chalmers and IH tractors in the 1970s and 1980s. We had at various times a G, a WD, a WD-45, a C, a 180, and a 190. They were sturdy and dependable machines. We did a great deal of work with them, and seldom had to work on them. I haven't restored one myself—yet—but I've maintained the tractors, and followed them at shows.

The first dealership we dealt with was J&R Implement in Albion, Wisconsin, known fondly by some customers as "Junk & Rubbish." One of my favorite memories, from age 5 or 6, is getting a homemade gravity box wagon to pull around, painted orange. My grandfather had made the wagon, and on the way home from Grandpa's we stopped at J&R to show it off. The folks at J&R promptly applied several Allis decals to the wagon—I think they must have raided a 200 decal set for the stickers. Another time, they had a big open house at J&Rs, with the Allis-Chalmers girls (with the thigh-high boots and miniskirts), but the movie they were to show of the West Allis factory had become brittle riding around in a cold car, and broke. I had always wanted to tour the Allis factory, but they had an age limit of 14. I could've gone in the narrow time frame between when I turned 14 and when the factory closed in 1985, but I didn't realize the closing was imminent and by the time I heard about it, it was done.

After J&R sold out, we dealt mainly with I. G. Hall in Johnstown, Wisconsin, with frequent trips also to Hanley's in Sun Prairie. They still maintain an impressive spare parts inventory. We never owned a tractor newer than the 180, although I would've liked to. J&R did the usual things to the injection pumps, so a four-plow tractor became a six-plow tractor anyway.

In 1985, Allis-Chalmers was undergoing a severe financial crisis, and sold its agricultural equipment line to Deutz. Deutz consequently shut down the line in West Allis, and imported tractors from Germany. Deutz changed the engines to air cooled, changed the orange to green, and blew what market share Allis had in tractors. The combines from Independence stayed pretty much the same, although the ag implement line was gone.

In 1990, the U.S. operations of Deutz-Allis were purchased by the American management, which formed a company called AGCO (Allis-Gleaner Corporation). The color was changed back to orange, and liquid-cooled engines made a comeback, with some engines being purchased from Navistar, the old International Harvester. Tractor production was resumed at the Independence plant for a few years in the 1990s, then moved to the Coldwater, Ohio, plant when AGCO purchased the White Farm Equipment Company. The ag downturn in the late 1990s, caused by the so-called Freedom-to-Farm Act, resulted in the Coldwater plant being closed down. The official closing date came while I was completing research for this book at the Independence engineering facility. All orange tractors will now come from overseas. At this time, the Allis name and orange color live on, although rumors suggest even that last vestige of the wonderful Allis heritage is at risk.

Acknowledgments

This book has one author and one photographer, yet many people helped put it all together. On the research side, thanks go Alan Jones, AGCO, for escorting me in the three weeks of research at Independence, helping me, answering the occasional dumb question, and pursuing at least one wild goose chase when the materials were right in front of me. Norm Meinert assisted with research materials, accompanied and helped me during one week in Independence, and shared much information. Norm Swinford provided some of the archival photographs in this book, and authored two wonderful Allis-Chalmers books that proved invaluable references. Jay Saleh, AGCO, assisted with running of various duplication machines, and a few other AGCO folks shared donuts and conversation, and in general made the weeks in Independence go faster. Lorna Jones of AGCO in Georgia got the book project approved. The staff at the Crossland Hotel in Independence took care of me as well. The people at Hanley's in Sun Prairie, Wisconsin, provided me with parts books and operator's and service manuals at cost. Lisa and Sherrie at the State Historical Society of Wisconsin helped in reproducing the other archival photographs in this book.

Thanks go to the following owners for the kind use of their tractors and themselves for photography: Wendel Bennett, E Thresherman's Special; David Ferguson, A; Larry and Edwin Karg, U, C with two-speed box, and RC; Norm Meinert, IU; Del Endres, UC; Dan Nelson, Unstyled WC High Crop; Paul Demmler, Styled WF, C all fuel, and G with loader; Aric Onsrud, WD; Dave Buttjer, WD-45 and WD-45 Diesel, CA; Don Paske, B; Charles Felger, IB; Ralph Doubek, Snow Tractor; Lorrie Kiefer, Gs

And at MBI Publishing Company, I'd like to thank Lee Klancher, Steve Hendrickson, Jane Mausser, and everyone else involved with this project.

Introduction: How To Use This Book

The purpose of this book is to describe Allis-Chalmers tractors as they appeared when they left the factory. For the conscientious restorer, it is a reference that covers what other restoration and service books do not, but it is not a comprehensive restoration book. You'll still need parts books, service manuals, and the like to finish the tractor and make it run right.

This book is split into several sections. The first chapter covers an introduction to the tractors and some general issues to consider in performing a restoration. It also covers the way Allis-Chalmers structured its information and its relevance to restoration.

The second chapter covers decals, whose importance is unique to tractor restoration. While cars usually have just a few exterior decals, tractors are loaded with warning, instruction, model information, and trademark decals that are the most noticeable part of a restoration. Most restorers and suppliers fail to realize that the decals varied considerably over the production life of a tractor, especially as service decal sets were produced. Service sets were often furnished with decals for newer tractors. While there's nothing that says you have to use the decals that came on your tractor when it was new, doing so maintains a little more history and gives restorers another point of interest when comparing tractors of the same model but different vintages.

The third chapter covers common parts used throughout Allis production, not just on one model. Oil gauges, for instance, were carried under one part number from the late 1930s on.

The next several chapters focus on the different Allis tractor families. The U family, for instance, consists of several tractors made with similar parts, including the U, IU, UC, and others. The chapters highlight the differences between the tractors, how the tractors changed over time, and describes the options and attachments.

The final chapter covers paint, wheels, and rims. Because these issues are often the most controversial part of a restoration, the chapter describes where the factory actually applied paint, and what wheels, tires, and rims were used in production. Along with their specifications, the coloration of wheels and rims often changed several times over the course of production, and various suppliers were involved.

Finally, serial numbers for tractors are given at the end of the book. Since the serial numbers of many changes were not recorded, or at least have not survived, the restorer will sometimes have to guess about what part of the year a tractor was built and carefully examine the tractor to determine whether a particular change applies.

ALLIS-CHALMERS

CHAPTER 1

Restoring an Allis-Chalmers

Allis-Chalmers was unlike any other company in the tractor business in the glory decades of the 1920s, 1930s, 1940s, and 1950s. It was much more than a farm equipment company—Allis operated divisions that ranged from flour mill equipment to nuclear power plants and fuel cells. The heavy capital equipment portion of the company was legendary for its accomplishments, producing huge pieces of equipment with microscopic accuracy. The technology that flowed from the engineering-heavy company washed over the farm equipment division, producing changes that shocked the equipment industry. Innovations in metallurgy and power transmission and a totally fresh approach to tractor configuration, combined with some highly experienced personnel trained in the other farm equipment companies, enabled Allis to briefly challenge for the lead of the farm equipment industry. Yet the wide diversity of the company also hurt the farm equipment division by denying it the investment necessary in the late 1930s to meet the demand for its products. The products Allis designed and built were so good that other companies were forced to mimic them. The other companies' larger dealership bases enabled them to overcome the Allis technology advantage to a degree, although in many cases the other companies were forced to pay Allis patent fees.

The early, large Allis-Chalmers tractors can be difficult to find parts and information for. Steel wheels aren't the easiest things to work with either. Yet a good restoration to original condition will make the tractor stand out among the rows of more common later tractors.

Restoration Philosophy

The beauty of the antique tractor hobby is that there is no judging at shows. In the car world, the most obsessive collectors bring cars to "prestige" shows in enclosed car carriers, then lay carpet from the ramp to the show area, so that the tires (with the stickers still on them) never hit the ground, and never have to be cleaned. The cars are never driven. For someone who grew up on a farm, the idea of treating a tractor like that is ludicrous. After all, tractors pulled manure spreaders. Yet some car shows are for viewing only, and attract vehicles in varying states of originality and restoration. With no judging, tractor collectors have the freedom to achieve the restoration standard that best suits them. If you want to bring your machine to the degree of originality specified here, great. If not, it may still be fun to know what has changed on your machine over the years, and what it might take to put it back to the way it came from the dealer.

Tractors in good, original condition should be preserved to document what they "really" looked like, as far as decal location, paint, etc., are concerned. Complete, clean, original machines grow rarer each year. At what point the wear and tear on a tractor makes restoration worthwhile is a decision for the owner alone. You must decide whether the wear on your tractor just gives it character or cries for help.

Sometimes, it becomes an obsession to restore tractors to factory condition. With farm tractors, however, other possibilities exist—along with more than one definition of originality. Dealers installed various pieces of equipment on the tractors they got from the factory. For example, LPG conversions were popular, and many tractors received LPG conversions

RIGHT
This is the front view of the WC Cane tractor supplied to Thompson, a manufacturer and distributor of sugar cane equipment. These layout drawings are from the files of AGCO. *Drawing courtesy of AGCO Corporation*

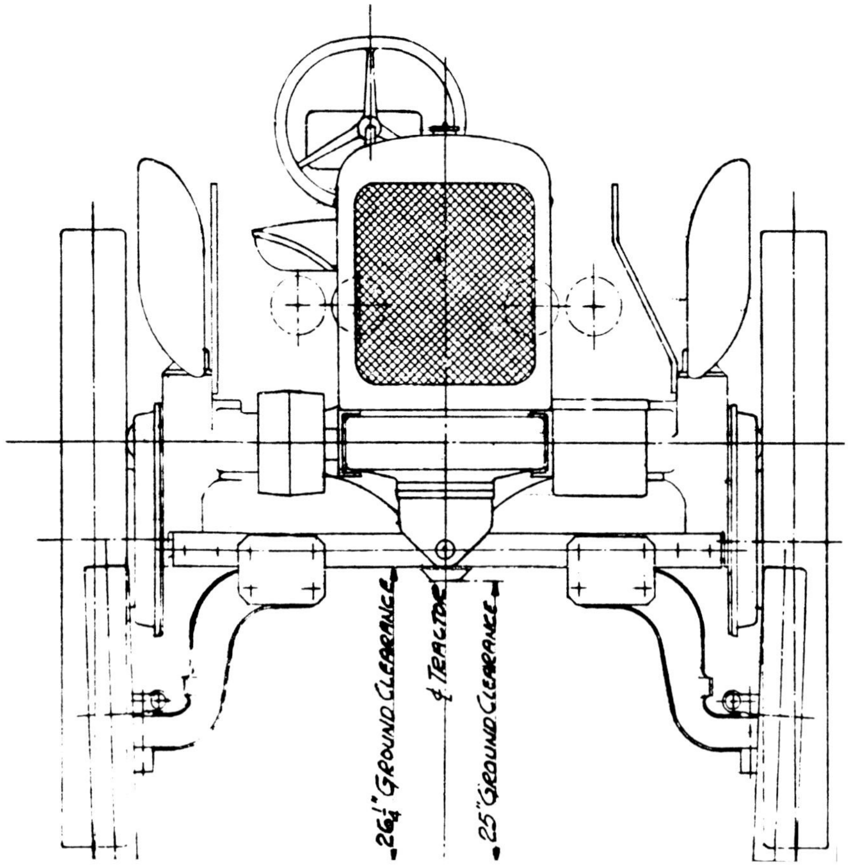

BELOW
This is, of course, the streamlined version of the WC. This version included 26-inch-diameter front wheels and 54-inch rear wheels. It would be interesting to see what Thompson put on the tractor after it received shipment. *Drawing courtesy of AGCO Corporation*

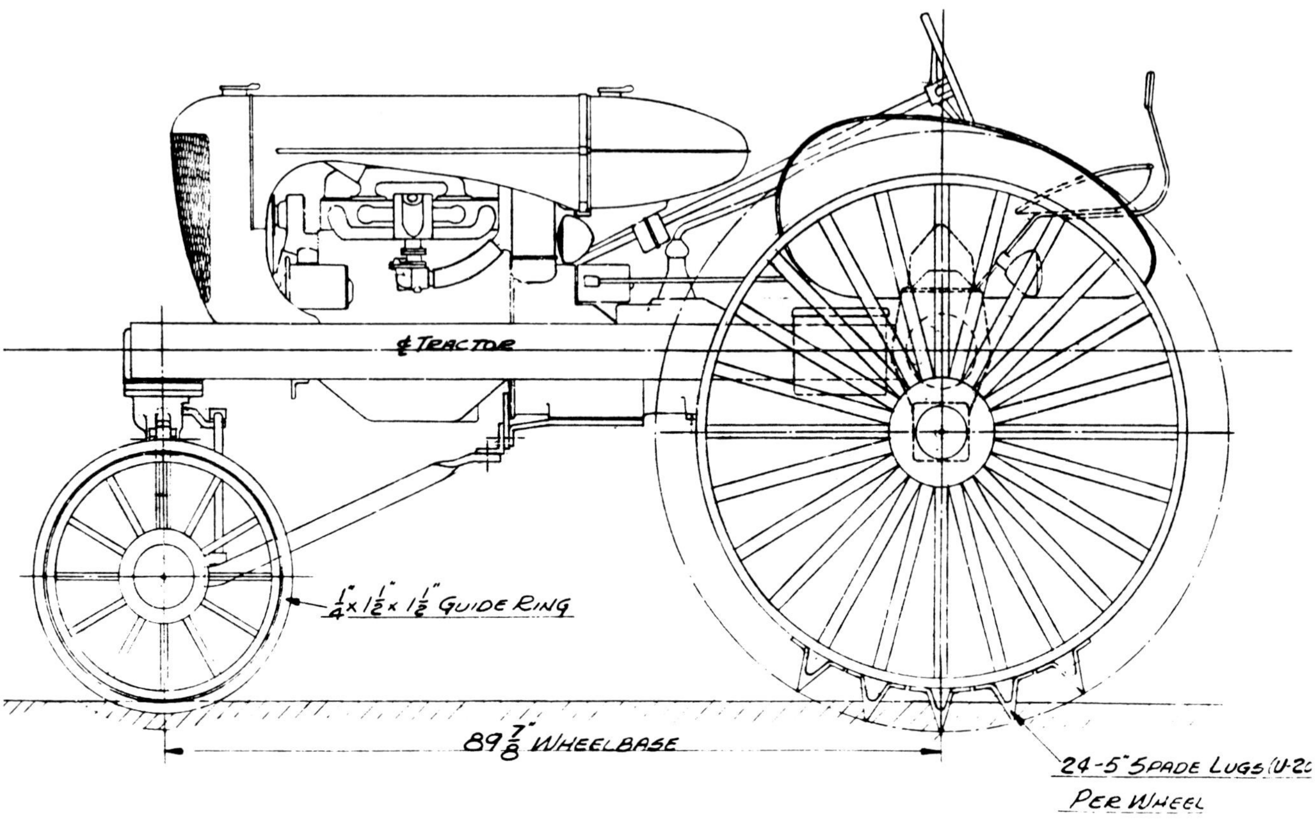

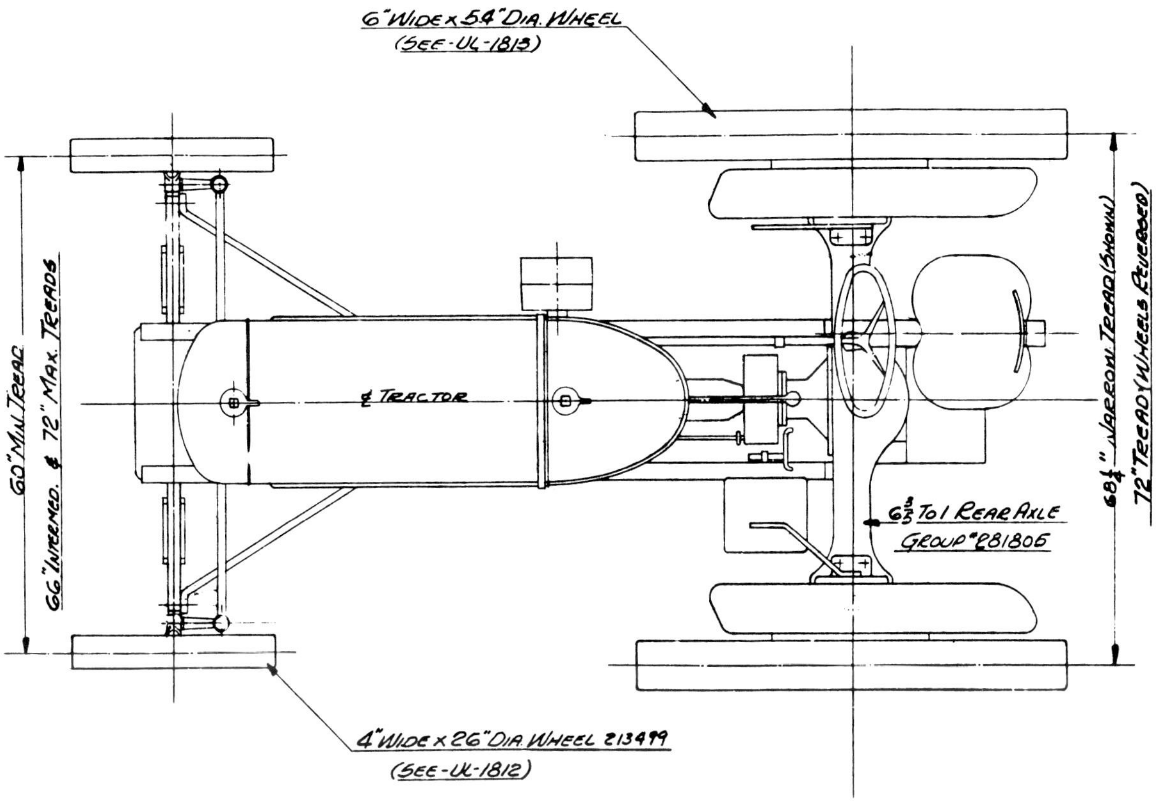

The top view of the WC Cane for Thompson (steel wheel version) gives the tread variation and some of the options. These layout drawings were intended for various purposes, including planning the shipping of these tractors. *Drawing courtesy of AGCO Corporation*

when they were brand new at the dealership. These tractors ran 40 years for their farmers, all of their lives, in that condition. Why remove that LPG conversion just because it wasn't "factory"? If it's your choice, that's fine, but LPG conversions were an important part of farm history, and they deserve to be preserved as well.

Sometimes other equipment was added by farmers or original equipment manufacturers (OEMs). The welder surely must rank as one of the top 10 inventions that changed agriculture. If a farmer added some brackets or made a homemade high crop conversion, the modification may be worth preserving. If a cane farmer took a cane tractor and put the gigantic "claw" cane loader on it, retaining that deserves consideration as well. The point is, "factory" isn't the only way to go.

"As Maintained" Restorations

Many farmers maintained their tractors pretty well. Routine maintenance items such as generators and starters got replaced, but farmers could and did get their tractors repainted and redecaled as well. Sometimes a dealership did the job for free, kicking it in as part of an engine overhaul. The tractor you remember driving as a child might have been changed before you were even born.

Decal sets as sold by Allis-Chalmers were not the same decals, either in appearance or number, as those put on a tractor originally. The decals were whatever was current when the decal set was produced, not what was current when the tractor was produced. Chapter Two shows what the service sets looked like.

In addition to decals, replaced components were often different from those originally on the tractor. Starters and generators may have been ordered from the dealership, or they could have come from the nearest auto parts store or a catalog like Central Tractor or Surplus Tractor Parts. These replacement parts were probably painted black, not orange. When Allis bought these parts, they were painted with the rest of the tractor, but when a farmer put on the new starter in the middle of planting season, the last thought on his mind was painting it orange. It worked just fine whatever it was painted, and it usually got left that way.

Sometimes parts were sent in to specialty shops for rebuilding, and in many cases, the parts came back painted black. In fact, this still happens today, so if you have a part to repair that is original color, be sure to tell the rebuilder not to paint it.

Safety

One restoration standard that can have no compromise is safety. While tractor safety is beyond the

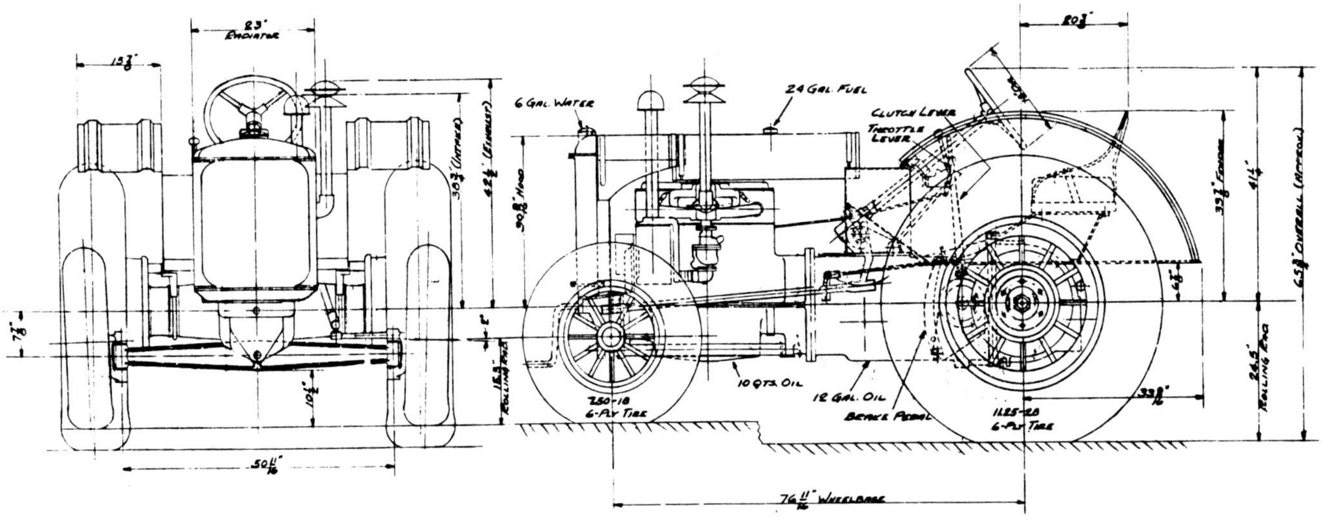

The front and side views of the UM tractor winch supplied to Cooper. Presumably, after Cooper got them, it would have the winches themselves added. The two different versions of this tractor were rated at 15 and 30 miles per hour. *Drawing courtesy of AGCO Corporation*

scope of this book, safety is critical if you are going to work with or show the tractor. Nobody likes to think that an accident can happen at a tractor show, but having worked with security at a tractor show for several years, I can report that tractor accidents do happen with scary frequency.

Mechanical reliability is critical if you are going to have the tractor out in public. Here are some critical areas:

1. **Brakes.** Brakes must be fully operational, with lots of pad. Brake pedals should not have side-to-side wear which can cause binding or allow a foot to slide between two pedals. Brake locks must work, and not be blocked by sheet metal or other parts. It's not fun seeing a six-year-old kid run over by a "parked" tractor with defective parking locks. I have.
2. **Fuel tank and cap.** The fuel tank and system must not leak. The ventilation holes on the cap must be open and not blocked by paint or rust.
3. **Exhaust manifold.** An exhaust manifold leak aimed at the fuel tank can heat the tank up, causing venting or fire, not an obvious thing, but it's happened.
4. **Steering.** Make sure the steering wheel isn't going to break off in your hands, especially if you're driving the tractor in road gear to the show. Make sure there is no excessive play in the steering box.
5. **Belt pulley.** Make sure all bolts and other hardware attaching the belt pulley are in good shape. If you are using the belt pulley, be sure that there are no cracks or chips. A runaway belt pulley can go for miles before it stops, or is stopped by something (make no mistake—this has happened at a show). If you're going to use the pulley, make sure it won't damage a belt. Broken belts can penetrate skin—this has happened too.
6. **Safety wires.** Make sure that all pieces of hardware that are supposed to have safety wires or lock nuts actually have them, and make sure all fasteners are torqued to the proper specifications. Flying cylinder heads can hurt.
7. **Rings, valves, muffler.** While the possibility of injury from these parts seems more remote, make sure piston rings, valves, and mufflers are in good shape. I've been bathed in junk and oil coming out of mufflers, and having some of this stuff get in an eye might result in an emergency room trip.
8. **Tires.** Make sure that tires are in good shape. A tire that blows can take someone's arm off, if it doesn't kill the victim outright. Split rims are notoriously dangerous. Make sure that lugs are fastened well on steel wheels. A flying lug can ruin someone's whole day.
9. **Wiring.** Old wiring can cause a fire, as can botched rewiring. A tractor fire is very exciting: on the farm, we had an Allis C that a previous owner had rewired for 12 volts. I was out cultivating one day, when I noticed flames coming out of the switch-on, which just happened to be located near my crotch. Very exciting.
10. **Seats.** Make sure the seat pan isn't corroded, the attaching hardware is in good shape, and that the springs aren't cracked. Seats can and do fall off, almost always when there's an operator in them. The tractor doesn't stop, and neither does the disc behind.
11. **Hydraulic system.** Make sure that hoses and other hydraulic system parts are in good condition.

Allis used a unique high-pressure system, for which maintenance is especially critical. The pressures are enough to break skin and cause real problems for an eye. A hydraulic oil shower is just plain unpleasant in any event.

12. **Lights.** If you are going to operate at night, your tractor's lighting system will need to be 100 percent operational.

13. **Protection systems**. Roll Over Protection Systems (ROPS) are roll cages that protect tractor operators during rollovers. People often buy old tractors to go mowing. They proceed on to hillsides, roll over, and get squashed. Other people soup their tractors up and go to antique tractor pulls. When the tractor does a backflip, the result is the same. ROPS are now available for many older tractors. Don't think that you can just "weld up" a roll bar—tractor roll bars can be rather finicky as far as materials and engineering, and doing it wrong can kill you. The transmission case can shatter during a roll, dousing you with hot oil, or you can be crushed between the fender and bars. Contact your county extension agent for more information about ROPS.

14. **Weights.** Make sure your tractor is weighted properly, especially if you are plowing.

15. **Clutches.** Make sure your clutch is fully functional. A nondisengaging clutch can be very dangerous, even though the tractor can still be operated. People have been killed on tractors with clutches that drag enough to pin them against a low overhang, strangling them as their throat contacts the steering wheel, even though they have the clutch pedal pushed to the stops.

Don't think that this list covers everything. Use your head, and every available resource, to make an honest evaluation of your tractor.

Other Information Sources

For people thinking about restoring a tractor, several different information sources should be found, begged, borrowed, stolen, or (preferably) purchased. These include:

1. **Parts books.** Parts books show many of the parts in the tractors, and later manuals show exploded views, which give a good idea of how to put a tractor back together. However, the only available parts manuals are often the latest revision, failing to show superseded parts. For Allis this is especially true—although a parts manual may show all of the parts used in producing the *last* tractor of a particular model, it will usually show only about half the different parts used on *all* tractors during the production run. The only changes

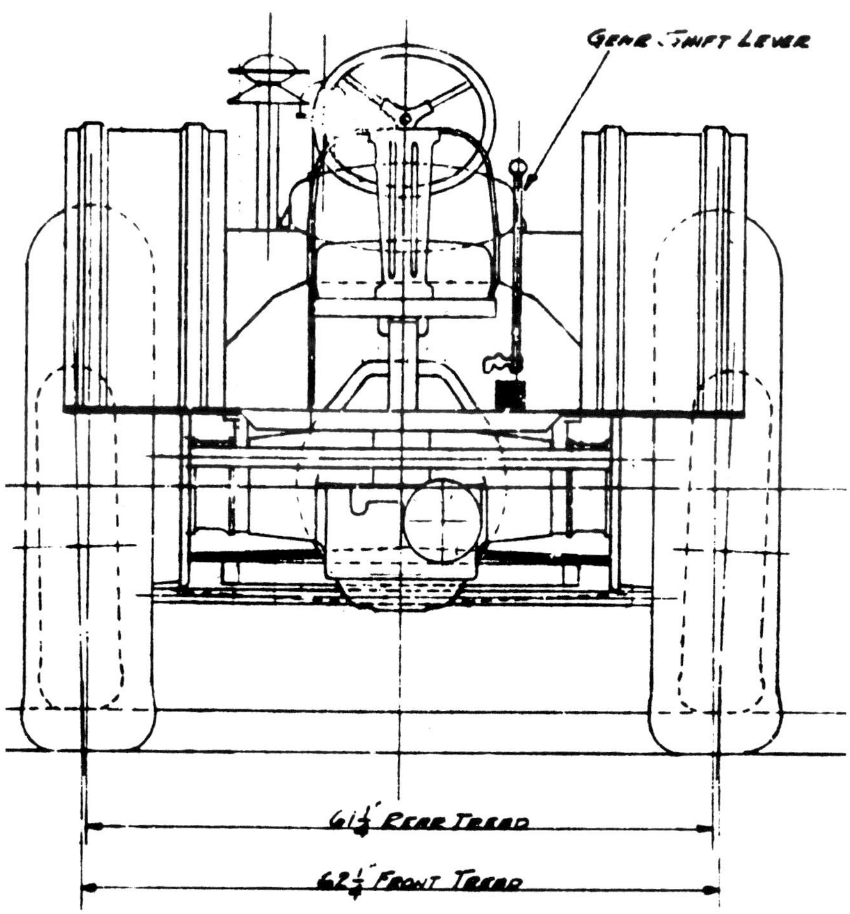

The Cooper UMs had an unusual gear shift, as seen here. The flying saucer muffler was common to the Model U tractors. *Drawing courtesy of AGCO Corporation*

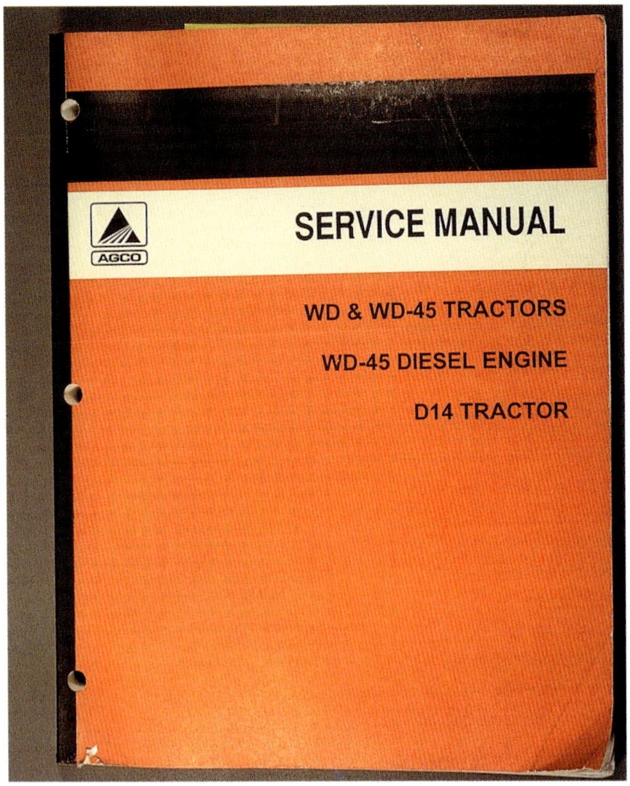

shown are usually those affecting the purchase of spare parts, and where those parts are available.

2. **Service Books.** Allis produced service books called *Knowledge Means Power* that included basic information and maintenance tips, as well as more detailed information about inspection and repair. Allis produced other manuals that were dedicated service manuals. In addition, I&T service manuals give actual disassembly and assembly procedures.
3. **Operator's Manuals.** These give operating procedures, as well as identifications and other information.

Parts Number Information

Allis-Chalmers used several different parts number sequences during its tractor production. For the most part, these numbers ran sequentially, and they are

If you're restoring an Allis, you should try to find as much original factory literature as possible. Shown is a service manual for WD and WD-45 tractors, along with a dealers parts catalog for B, IB, and C tractors, and a "Knowledge Means Power" service book for the Model A. These can be found through literature dealers and at swap meets.

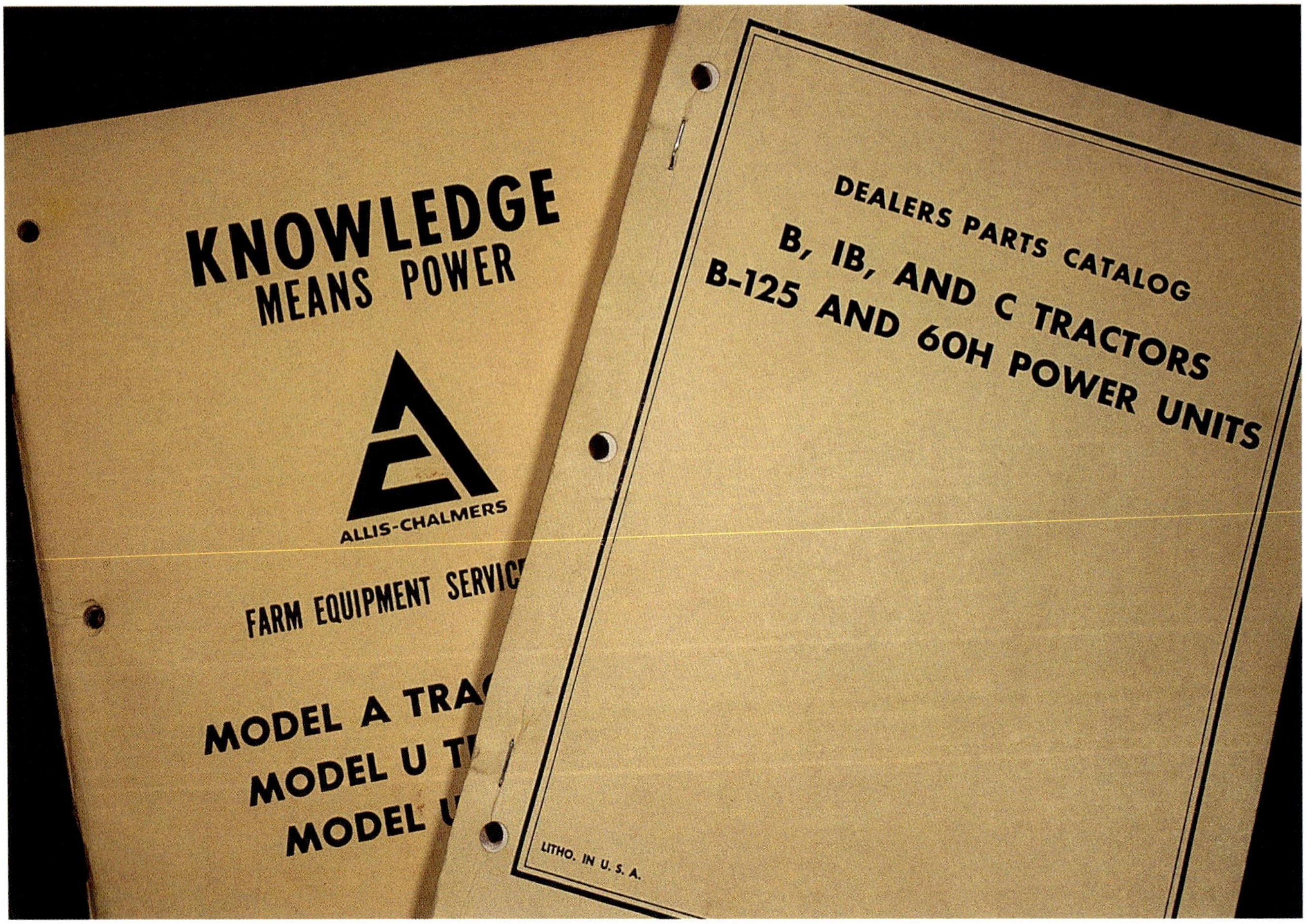

simple numbers, containing no information about revision or usage.

The first parts numbers Allis assigned to West Allis–produced tractors were the AM prefixes (example: AM-1200), dating back to the tractor truck, rototiller, and 9-18, produced (or experimented with) before 1920. In the 1920s, Allis began producing parts with the E prefix (example E-1200), which matched the E models then produced. In the late 1920s and early 1930s, Allis began using U prefix parts numbers, again for the U model tractors, although a few A parts may have received the U. The E and U parts sequences contain parts used on tractors other than the E and U models.

In the mid-1930s, Allis abandoned the use of letter prefixes and began using numbers, starting where the U parts numbers stopped. In 1962 the Allis engineering department converted the U numbers into the 200000 numbers by adding 2 and 0 in front of the four-digit U number. Thus, U-1 was converted into 200001. Approximately the first four thousand 200000 parts numbers are converted U numbers.

Three sequences were contained in the 200000 series parts numbers. The numbers starting with 200001 were parts numbers. The numbers starting with 280000 were called "groups." These were major subassemblies made up of various parts. The numbers starting with 290000 were complete tractors—base tractors and tractors with various option groups. The 280000 numbers are encountered by collectors as titles of drawings in some of the service manuals, while the 290000 numbers (known as "catalog" numbers by collectors) are located in dealer price catalogs and sales invoices, if the collector should be so lucky as to have either of these.

While the collector and enthusiast might think that Allis built only a few different tractor models over the years 1934–1957, the Allis catalog number system, using a separate number for each major variation, resulted in just over 800 different "tractors" including power units, replacement engines, a few experimental units such as the F, D, H, and B-2, and some military items, including snow tractors and scout cars. These variations were produced with 30,000 different major parts, and untold thousands of different small "standard parts" (nuts and bolts).

Other factories and lines used different blocks of numbers. Model Gs produced in Gadsden, Alabama, used 800000 series parts numbers, with group numbers in the 845000s, although before the transfer of the G project to Gadsden, the G had West Allis 200000 numbers. Parts in the 900000 series were parts that were used companywide, including common hardware such as nuts and bolts, but also paint and a few decals. Some decals that originally had West Allis 200000 numbers were renumbered into the 900000 series and used companywide. Harvey Works (the ex-Buda Company purchased by Allis-Chalmers) supplied the diesel engines in the WD-45 Diesel tractors, and also had different parts numbers.

Many parts were purchased from outside suppliers, a practice Allis used heavily. If the part was common to other manufacturers, the outside supplier's number will be on it, although Allis designated an Allis part number to it for record keeping, parts ordering, and inventory purposes. A part failing to have an Allis number, or having another manufacturer's name on it, doesn't mean it's not original equipment to the tractor. The vast variety of rims and wheels that Allis offered, often for special crops or conditions, means that a variety of manufacturers will be represented, and equipment will be seen that is not in the parts books.

It seems that only about half the parts designed and held in the blueprint files of Allis-Chalmers (now AGCO, of course) actually show up in the parts books. Most of the missing parts numbers are for specialty tractors not commonly available, superseded parts on which production was discontinued, specialty equipment produced for other OEMs, or other manufacturers who purchased Allis tractors as the base for their own equipment. There can be other reasons as well.

In general, Allis produced a lot of variability. Instead of being just one tractor with a few options, Allis' system created several different tractors, all slightly different. The trail in researching this book has been very difficult due to Allis' engineering documentation system. Since a few critical pieces that tie the massive amounts of information together seem to be missing, this book should be taken as a general guide, rather than a comprehensive "bible."

Changes and Dates Used in This Book

When a date is given in this book for a product change, it is important to note that unless accompanied by a serial number, it is the date the change was authorized by the engineering department, usually by the chief engineer's signature. Changes did not take hold instantaneously, but were delayed by the need to produce new tooling and stock and, in many cases, by the need to go through old parts first. Changes for safety and major reliability issues were usually made as quickly as possible, with the stock of old parts scrapped; but when parts were reengineered to reduce cost, or for other noncritical reasons, the old stock was used up first. Time lapses between approval and implementation could extend from a month to a year, although on average, the changes were more likely to be in the range of a few months. In certain cases, it is also possible that changes occurred before the authorization.

TO ADD WEIGHT. ENOUGH ANTI-FREEZE SOLUTION PARTLY BE ADDED TO MAINTAIN A PRESSURE AIR MUST ALSO OF 28 POUNDS IN FRONT AND 12 POUNDS IN REAR TIRES.
POWER-CRATER ENGINE
SNAP-COUP
ALLIS-CHALMERS
A-C
PARTS
ENGINEERED RIGHT
MANUFACTURED RIGHT
TO PERFORM RIGHT
226503
ALLIS-CHALMERS
MILWAUKEE 1, U.S.A.
WD
45
TRACTION
POWER IMPLEMENT
CONTROL

CHAPTER 2

Decals

Allis used a variety of decals that changed over the years. Generally, decal placement wasn't as rigidly specified as with some other tractor companies, although in some cases placement was specified to within fractions of an inch.

Service decal sets were sold to dealers and farmers when tractors were repainted. These sets did not necessarily represent the decals the tractor had when it was produced. Decals in the service sets were current when the sets were produced, so later decal styles were often substituted for earlier ones. And certain decals were not provided with service decal sets. Typical decals not included with sets were patent decals and certain instructional decals. Decals provided with other spare parts were also not included.

The service decal sets came in sealed cardboard tubes marked "Allis-Chalmers Parts." The decals usually had an Allis part number on the back, with a dash and then another number representing the revision. The decals themselves were not made by Allis-Chalmers, but were supplied by several outside vendors.

Unfortunately, there are some suppliers to the antique tractor hobby who are making flagrantly inaccurate decal sets today. They are advertised as "historical" or "aesthetically correct" in an attempt to evade AGCO product licensing. These decal sets include WC or C model symbol decals that look like nothing Allis ever used. These decal producers, whose descriptive advertisements seem to make every attempt to confuse the purchaser, are, in this author's opinion, unethical.

The decal sets Allis-Chalmers sold through its dealerships from the 1940s through the 1960s came in tubes like these. This is a complete replacement set for the WD and WD-45 tractors.

Other vendors are licensed by AGCO, or at least make an effort to produce an accurate reproduction. In the hobby, the most respected decal vendor is Lyle Wacker. Wacker doesn't carry all the revisions you will see in this book, or that Allis produced, but his decals are closer than anyone else's.

Decal Placement

Allis-Chalmers never produced comprehensive decal placement charts or complete specification lists showing which decals were used in production. Decal placement was sometimes noted on the parts that received the decals, sometimes with informal drawings, sometimes by word of mouth in the factory, and sometimes not at all. Decal listings were included with the parts on which the decals were placed, but finding all the decals and placements has proved to be very challenging. Some drawings were just plain missing, while other parts lists included multiple drawings of the same group, without information describing which decals were used where. I've tried to cover all the decals, but may have missed a few, and there may have been variations in placement that existing records don't document.

Production Decals for Models E and A

Apparently, Allis never produced a decal service set for either the 25-40 or the A. However, the following decals have been identified as having been used on the tractors during production. There may be others as well.

Decal, fuel tank end, Models A and E (Allis-Chalmers diamond) 915644
Oil filter decals Various
Hood, Allis, Model A 915677
Decal, instruction, transmission case, Model A Serial Number 26682
and after (no drawing available) 209522
PTO warning decal (on As at least after February 1939) 210300
Decal, rear wheel rim (air tires), Model A 208663
Patent decals (probably Model A only)

Production Decals for Models U and UC

Allis-Chalmers apparently never created a decal service set for the U or the UC. The following decals have been identified as being used on the U and UC tractors. There may be others as well.

Water drain 206289
Purolator oil filter. The drawing for the oil filters shows the decal; the decal was from Purolator itself.
Allis oil filter decals Various
AC diamond on back of tank 915644
Clutch lube decal
PTO decal (after February 1939) 210300
Tire instruction decal
Patent decals
Shift pattern decal

Production and Service Decals for Models B, IB, and C

The first section is for decals known to have been used in production. The second section is the decal service sets listing. The production decals were used at some point in production—check the individual decal sections for more information.

Decal, drain cock 206289
Wiring diagram decals (parts manual refers to two wiring charts for magneto and distributor ignitions) 212227
Battery decal 210090
Allis Several different—see listing
Air cleaner instruction 209263
Air cleaner oil level decal (not located) 209264
Radiator shutter opening decal B model decal (see text for effective date) 225748
Instruction decal for clutch lubrication
PTO guard warning 210300
Front tire instruction 208663
Rear tire instruction 208663
Oil filter decal—can type 224973
Hydraulic ram 213291
Patent decal 211877

Service Decal Set for Model B, Part Number 213322 (also used for C)

2 Decals "Allis-Chalmers" 915677
1 Decal, battery instruction 210090
1 Decal, wiring diagram 212227
1 Decal, cylinder block drain 206289
1 Decal, radiator shutter 213161
1 Decal, clutch lubrication 208506
1 Decal, air cleaner oil level 209264
1 Decal, air cleaner instruction 209263
1 Decal, model designation 225748
(Note that this was the "B" decal—Allis never made a decal for the "C." If you ordered a set for the "C," you got a "B" decal.)
1 Decal, oil filter can 224973
1 Decal, wheel rim 208663

Production Decals for Model CA

Model decal 225716
Hydraulic pump (See hydraulic pump listing)
Air cleaner At least two different—see listing
Allis decal See listing
PTO decal 210300
Tire inflation decal
Decal, battery instruction 210090
Hydraulic ram decal
Hydraulic control hand lever decal See listing
Cylinder block drain
Patent decal 213538
(possibly others)

Service Decal Set for Model CA Tractors with Magneto Ignition 213322

There was another decal set produced for distributor ignition tractors, with a slightly different wiring instruction, as well as a battery filling decal. This set was eliminated in 1963.

Service Decal Set for Model CA Tractors with Battery Ignition 225338

This decal service setwas produced for repainting B, C, and CA tractors. This set was used well into the 1960s, if not later. Apparently, Decal Set 213322 was eliminated in 1963, and combined into 225338.

2 Decals "Allis-Chalmers" 915677
1 Decal, battery instruction 210090
1 Decal, wiring diagram 225333 or 212227
Combined previous set
1 Decal, cylinder block drain
1 Decal, radiator shutter 213161
1 Decal, air cleaner instruction 209264 or 3?
1 Decal, oil filter can
1 Decal, hydraulic pump instruction 228417
1 Decal, hydraulic pump control hand lever 223739
1 Decal, wheel rim 208663
1 Decal, PTO warning 210300
2 Decals, model designation "CA" 225716
1 Decal, torque tube cover 208506
1 Decal, Snap Coupler 226684

Production Decals for Models WD and WD-45

Allis-Chalmers decal	Several, see listings below
1 Decal, battery instruction	
1 Decal, air cleaner	
1 Decal, radiator shutter	See listings
1 Decal, steering shaft universal joint	
1 Decal, wheel rim	
1 Decal, pump control hand lever	
WD-45 decal	226502
WD decal	225749
Snap Coupler designation	
1 Decal, engine designation (Power Crater)	
Block cooling drain	206289
Oil filter element	See listing
Diesel decal	227070
Hydraulic pump decal (on battery box)	Several, see listing
Transport valve decal	227310

Production Decals for Models WC and WF

Water instruction	
Allis decal	915677
Air cleaner instruction	226229
Air cleaner instruction—varies by air cleaner	see listing
Radiator shutter or radiator curtain instruction decal	see listing
Decal Allis, on frame rails	915677
Tire instruction decal	208663
Steering gear lube	208543
Battery instructions	
Oil filter element	See listing
Wiring diagram	

Service Decal Set for Model WC 213321

Decal Set 213321 was created January 15, 1942, to provide decals for repainted WC tractors. This decal set was apparently kept current by Allis-Chalmers until 1963. After this, apparently Decal Set 226503 (the WD and WD-45 decal set) was used for service.

2 Decals "Allis-Chalmers"	(See Note 1)
1 Instruction decal (for filling battery)	210090
1 Engine block drain decal	206289
1 Wiring diagram decal	208401
1 Instruction decal (air cleaner)	(See Note 2)
1 Decal (radiator)	213161
1 Decal (steering universal joint)	208543
1 Container (the tube the set came in!)	912579
1 Decal (oil filter can)	224973 (See Note 3)
1 Decal (wheel rim)	208663 (See Note 4)
1 Decal-patent	228892 (See Note 5)

Note 1. Original decal was 207378. This was changed to 218140 on January 16, 1942, indicating that the previous number was probably erroneous and no sets were created with 207378. The trade name decal was again changed to 220165 on September 22, 1944. On February 1, 1946, the decal was changed to 913677.

Note 2. Original decal was 209517. Changed to 226229 in July of 1950.

Note 3. Oil filter can decal 224973 added April 7, 1950.

Note 4. Wheel rim decal 208663 added August 4, 1950.

Note 5. There were changes made, but the change notice was missing.

Production Decals for Model G

The group drawings for the Allis G were not found, and determining the actual decals used on production Gs was difficult. Presumably, most of the decals were included in the decal service set, with the addition of the patent decal, hydraulic ram decal, and possibly others.

Service Decal Set for Model G 800138

1 Decal, air cleaner	209263
1 Decal, gear shift	800104
1 Decal, battery instructions	210090
1 Decal, wiring diagram (front of battery box)	800106
2 Decals, Allis-Chalmers	915676, or 338338, or 924396

Decal 800887 was added in late April 1950 and deleted August 21, 1952. I could not find documentation referring to this decal.

(Hydraulic ram decal, not in set, says "top"; was included on new rams.)

Decal Descriptions

Model Symbol Decals

Allis did not use model symbol decals before 1951. The model symbol decal drawings were made in June 1951, and included the B, CA, and WD. The decals had a clear background, black letters with a 1/16-inch-wide white border, a 1/16-inch rectangular black border around the decal, and a 1/32-inch white border around that. The decals were 2 1/16 inches high, but varied in width with the model. It appears there was never a "G" decal.

The WD-45 decal and the "Diesel" decals were, of course, available when their tractors were available, and were done in similar fashion to the earlier model symbol decals in colors and bordering. All lettering was black, as were borders. Backgrounds were clear, and all decals had "Printed in U.S.A." located on the face.

The B and CA decals had their back edge located 1/4 inch in front of the sheet-metal bead, with the centerline of the decal located on the centerline of the attaching bolt located just to the rear of the bead.

The WD model decal was located 1/4 inch in front of the bead, with the bottom of the decal located on the centerline of the lower attaching bolt just behind the bottom of the bead.

The WD decal, 225749 revision 2.

The WD-45 decal, 226502 revision 2.

RIGHT
The location of the B and CA model decals. The centerline of the B is on the centerline of the nut. The right edge of the decal is 1/4 inch to the right of the seam. *Drawing courtesy of AGCO Corporation*

"Allis-Chalmers" Decal

After the model symbol decals, changes in the trade name decals on the side of the tractor make up the greatest subject of debate. The decals change as follows:

U-2756 was the frame decal for the WC, and was drawn before February 1934, when it was redrawn to conform to the original supplier's sample. The decal used "A" and "S" at the ends that were longer than the other letters. The letters were deep blue, surrounded by a white outline (except the "Made in USA"). The

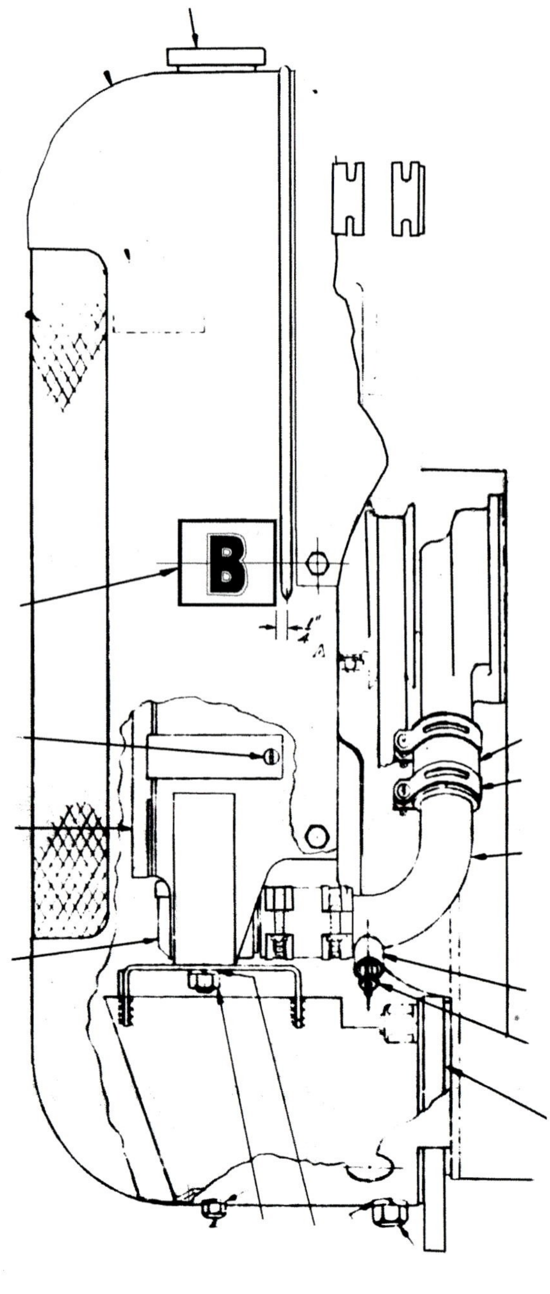

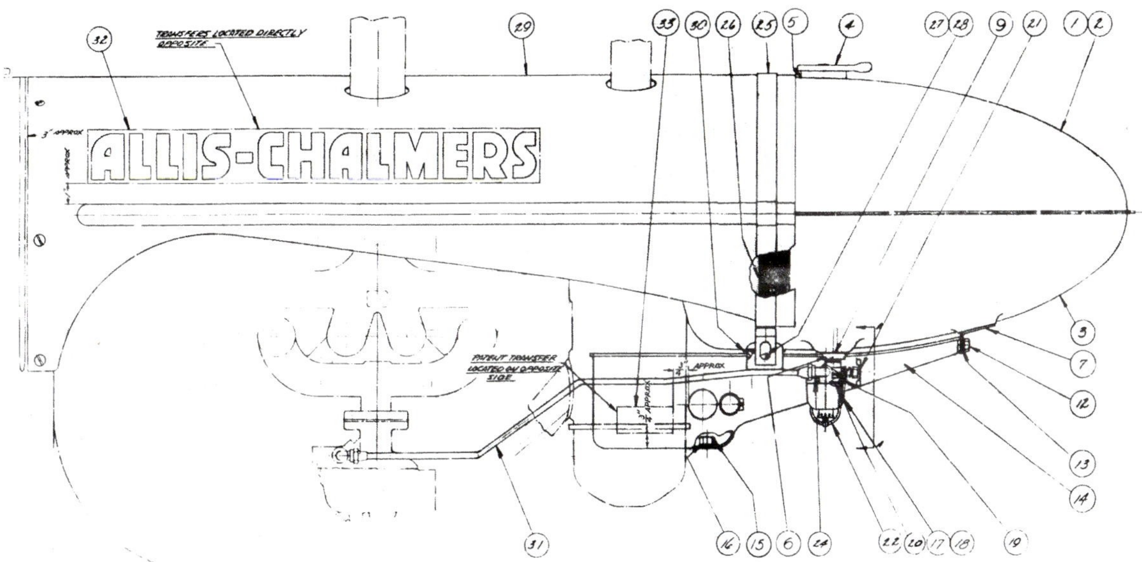

This places two decals on the WD and WD-45 tractors. The trade name decal's bottom edge is approximately 1 inch above the seam, and the left edge is approximately 3 inches to the right of the front hood seam. The right side is exactly opposite the left decal. The patent decal is located on the right side. The rear edge of the decal is 3/4 inch in front of the larger hole, and the bottom edge of the decal is 3/4 inch above the bottom of the fuel tank bracket. *Drawing courtesy of AGCO Corporation*

decal was surrounded by a deep blue border on the outside, with an inner white stripe. The width and height of the decal differed slightly among manufacturers (there were three and possibly four), as did the background of the decal, which was either clear or Persian Orange. WC decals before early 1934 used black lettering.

Decal 207378 was applied to the Allis B and C series tractors. The lettering was identical to U-2756, except that the white and blue borders around the outside of the decal (not the lettering) were omitted, resulting in a narrower decal. In fact, the parts drawing for 207378 has a note that the decal can be made from U-2756, which essentially would have required a paper cutter to trim off the borders and some of the background.

Decal 218140 was applied to the B and C family of tractors starting apparently in January 1944. This decal has the long "A" and long "S," but changed the color of the letters to black lacquer, and omitted the border around the letters. A clear background was used, the lettering was slightly smaller, and the overall dimensions of the decal changed slightly.

Decal 224841 was applied to WDs, and was first drawn in January 1950. This decal has a clear background, black letters, and a 5/64-inch-wide border around the Allis-Chalmers lettering and dash. The "Made in USA" has no border. The "A" and "S" are the same size as the rest of the lettering.

Locations of the "Allis-Chalmers" decals vary on the streamline tractors. On all low-octane fuel tractors with auxiliary gas tanks, the decal always lines up to have the stud for the gas tank on the right side of the hood pass through the hyphen in "Allis-Chalmers." The left-side decal is lined up with the right-side decal.

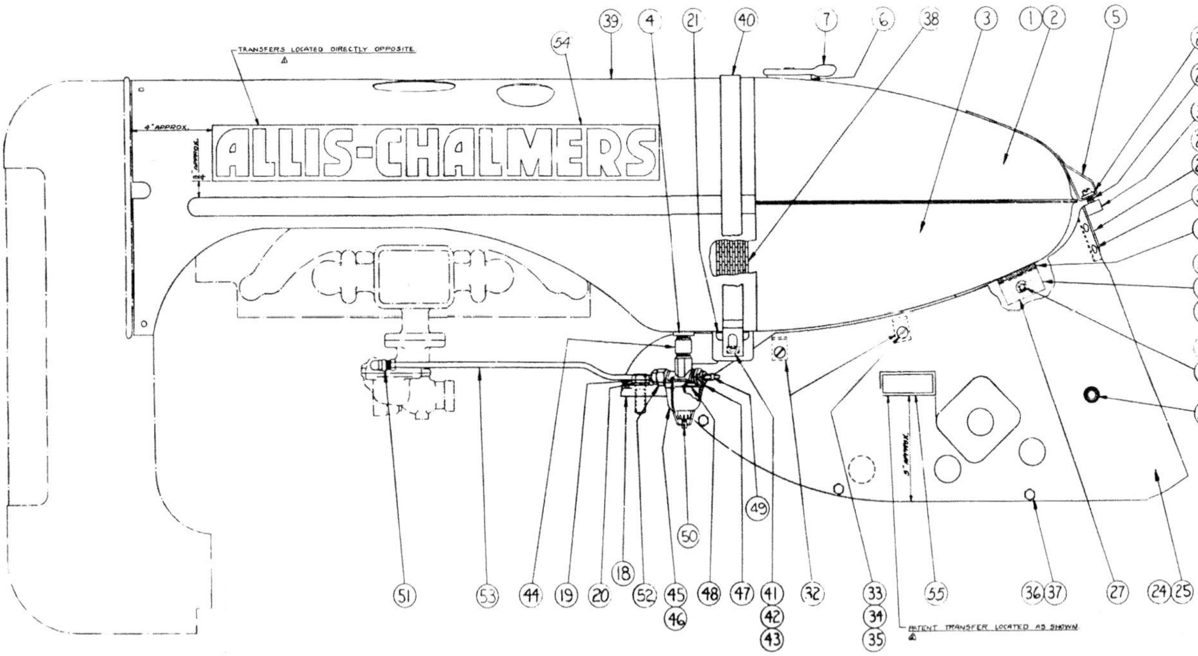

This locates several decals on the IB tractor. The trade name decal is 4 inches to the rear of the front edge of the hood and 3/4 inch above the seam. The patent decal is located approximately 5 inches above the bottom of the shroud. *Drawing courtesy of AGCO Corporation*

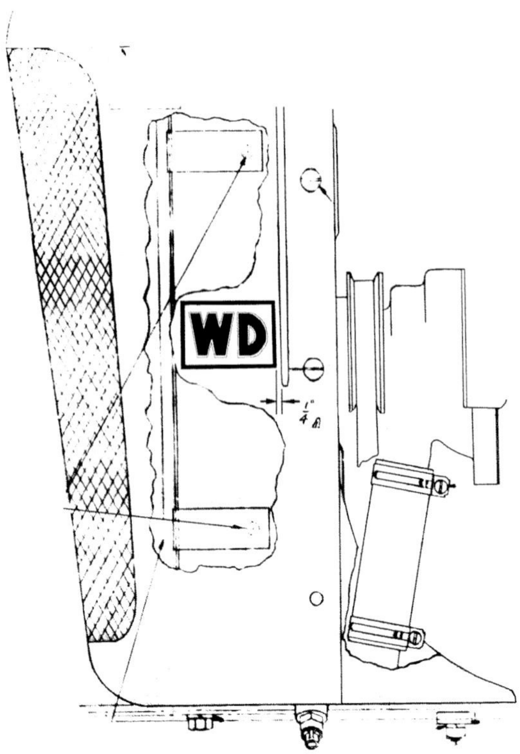

The location of the model identification logo on the WD. The decal's bottom edge lines up with the centerline of the bottom bolt, and the right edge of the decal is located 1/4 inch forward of the seam. *Drawing courtesy of AGCO Corporation*

On nonkerosene tractors, the positions vary with the model of tractor. On WD gasoline, WD-45 gasoline, and WD-45 Diesel tractors, the front edge of the left "Allis-Chalmers" decal is approximately 1 inch above the top edge of the beading, and 3 inches to the right of the front edge of the hood sheet. The right-side decal is directly opposite.

On B and IB tractors, the front edge of the decal is located 4 inches behind the front edge of the hood, and 3/4 inch above the hood beading.

In later parts books, the 200000 series numbers for the decals are replaced by 900000 series numbers. The 900000 decals have black lettering with white borders.

Allis Service Decals

Part Number	Height (inches)	Length (inches)
915675	7/8	8 3/4
915676	1 3/16	11 7/8
915677	2 1/4	22
915678	2 15/16	29 1/4

Fuel Tank Decals

These decals go on the flat rear surface of fuel tanks of unstreamlined tractors. The most famous decal was, of course, the Allis-Chalmers Diamond that appeared on the E, U, UC, WC, and WF tractors. This decal was centered on the rear panel of the tank.

Decal U-639 (200639) was used on United tractors, and consisted of the United Shield. The "United" was in black letters on a white outline, surrounded by

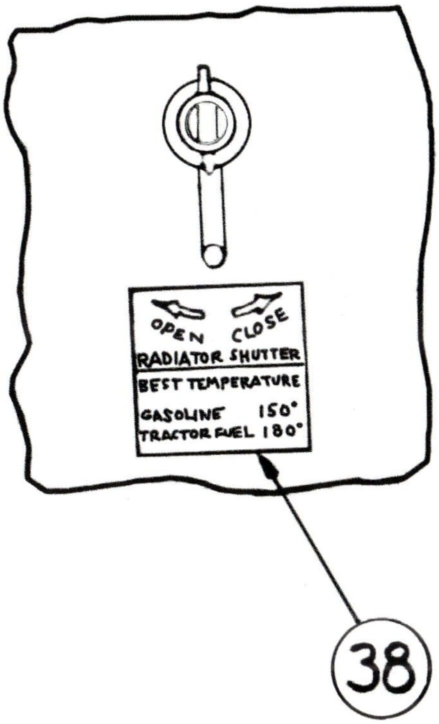

The location of the radiator crank instruction decal on B and C tractors. There are no dimensions on the drawing. *Drawing courtesy of AGCO Corporation*

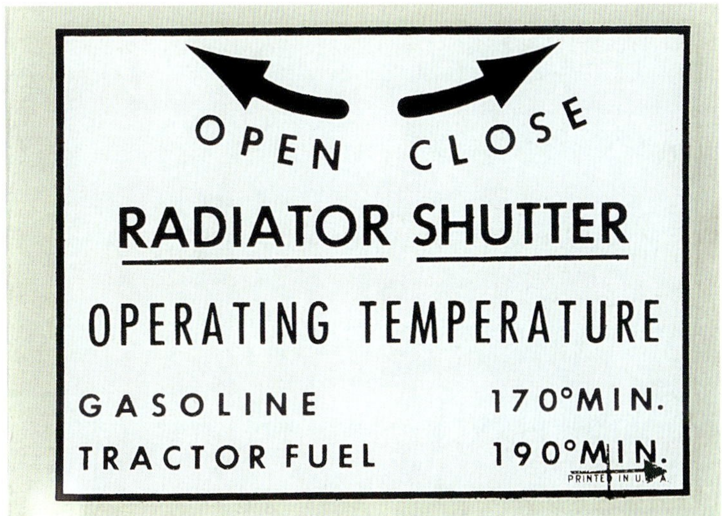

The Radiator Shutter instruction decal 213161 revision 4.

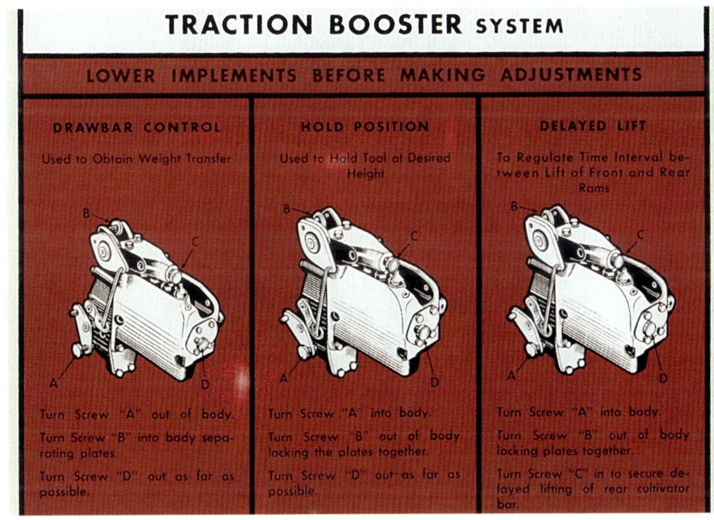

The hydraulic pump instruction decal, 228412 revision 3.

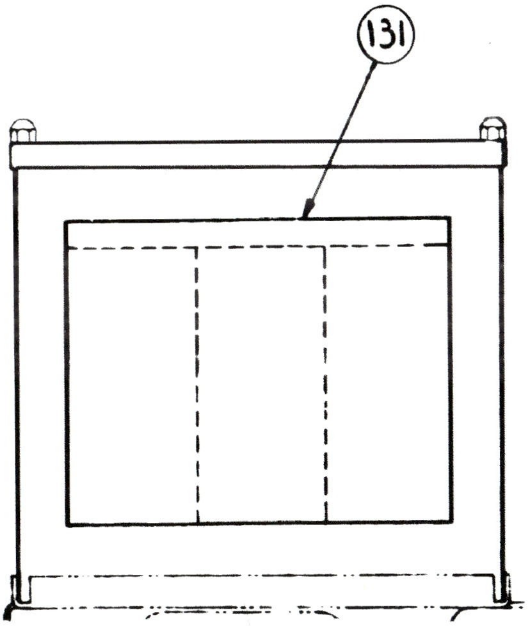

The location of the hydraulic pump operation decal battery box on the WD and WD-45 series tractors. *Drawing courtesy of AGCO Corporation*

Persian Orange, a black outline, a white space, and then another black outline. This decal was to be used on all tractors sold by United Tractor Dealers. It is not known how long this decal was used.

U-838 was used on Cockshutt tractors built by Allis, probably Es. It is an oval-shaped decal that reads "MFD. FOR and SOLD BY COCKSHUTT PLOW CO." The decal has a tan background with red lettering surrounded by thin black borders.

Radiator Decals

There were two types of radiator screen decals. The first decal, used on the W tractors, was for the radiator curtain. It told the operator when to use the curtain to keep engine temperatures high. The first W tractor decal was U-3348. In February 1939, that decal was replaced by the less-wordy 209333, which was replaced by 209516 in late March 1939.

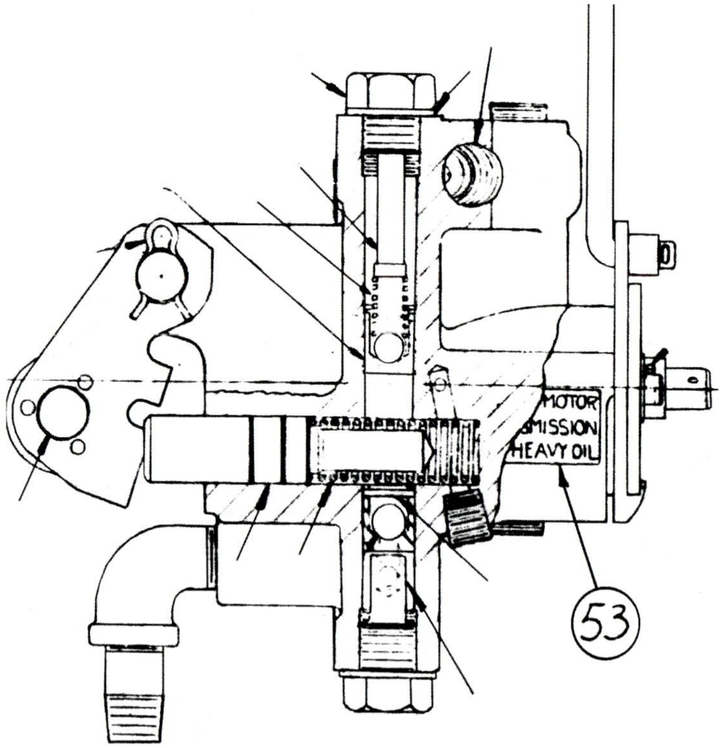

The number 53 refers to the hydraulic pump instruction decal, which is located on the pump itself on the B and C. There are no dimensions on the drawing. *Drawing courtesy of AGCO Corporation*

Decal 203348's text reads, "Cover the radiator when burning kerosene, distillate, or tractor fuel, sufficiently to maintain water temperature just below boiling point. A cold engine produces excessive crankcase oil dilution."

Decal 209333's text reads, "Cover radiator completely when starting. Open curtain when necessary to avoid boiling." Decal 209516 had the same text, but on three lines instead of four, with completely different dimensions.

All had black printing on a white background. The first two decals were made of waterproof paper on a

The "Power Crater Engine" decal, 226289 revision 1. Parts of this decal are transparent; see the description for detail.

BELOW: The dotted lines, part 28, show where the hydraulic pump decal is located on the left fender of the CA. *Drawing courtesy of AGCO Corporation*

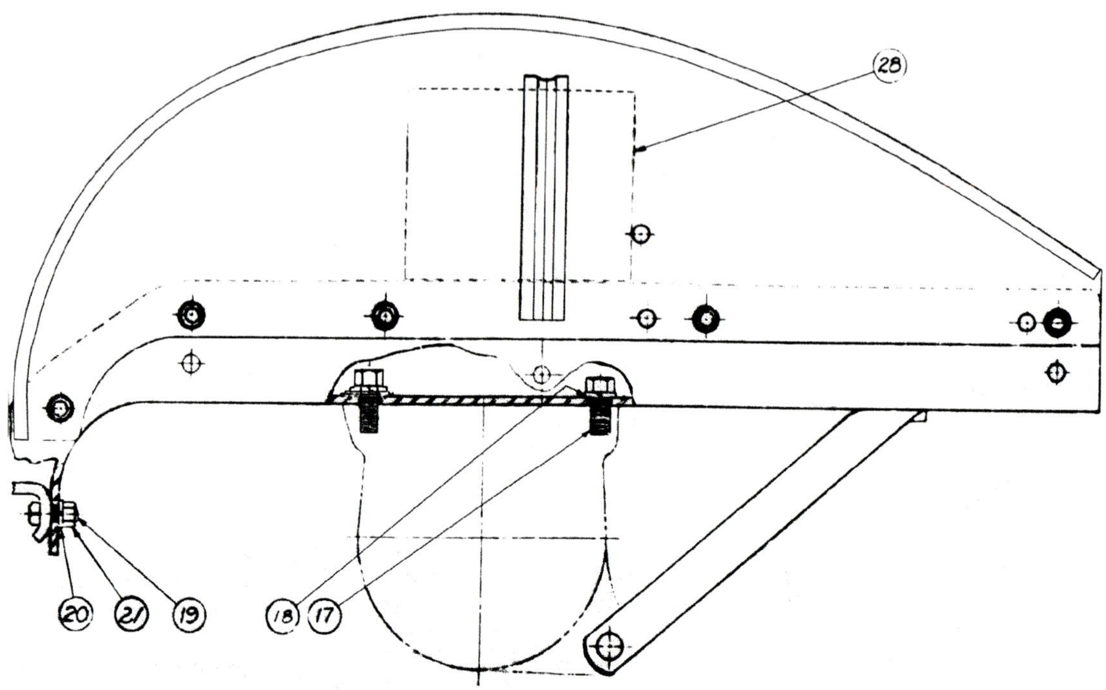

cloth background, while the last was of traditional decal material. The 25-40s, As, and Us had the decal located on the curtain itself, on the lower right-hand corner looking at the front of the tractor. The Bs that had radiator curtains also used 209516. These decals were located on the hood below the thermal gauge from the operator's viewpoint.

The second decal type was for the crank-type radiator shutter. These decals were also used on the W, B, and C tractors that used these shutters. The decal giving cranking instructions was located under the shutter crank. The decal was originally 212279, but was changed to 213161, drawn in late October 1941. Decal 213161 went through a major change in December 1941, when the decal was enlarged and the operating temperatures changed. Gasoline went from 150 to 170 degrees, and tractor fuel went from 180 degrees to 190 degrees. The wording "best temperature" was changed to "operating temperature."

The shutter decal on the B, C, and CA was positioned so that the top of the decal was just under the top edge of the radiator screen and just slightly to the rear. On the WC, WD, and WD-45 tractors, the decal centerline was aligned with the crank centerline, with the middle of the decal located where the crank reaches its bottom point.

Hydraulic Pump Decals

The hydraulic pump on the CA, WD and WD-45 series tractors required an instructional decal on

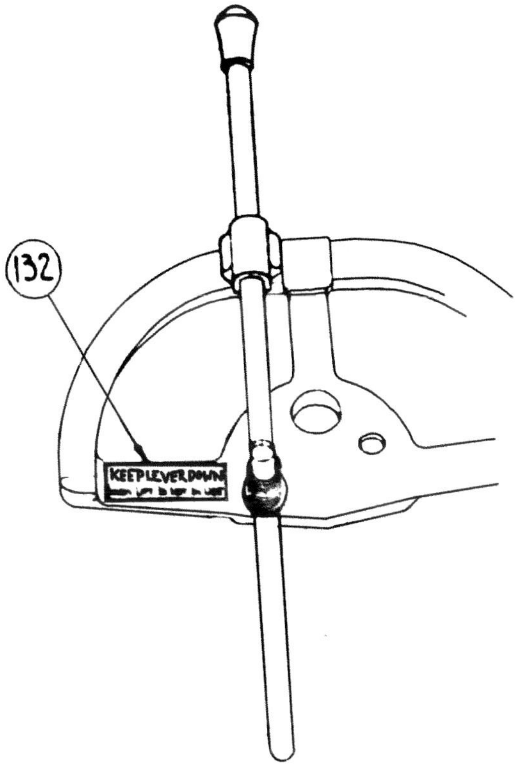

The location of the hydraulic lever instruction decal for the WD and WD-45. The decal says, "keep lever down." *Drawing courtesy of AGCO Corporation*

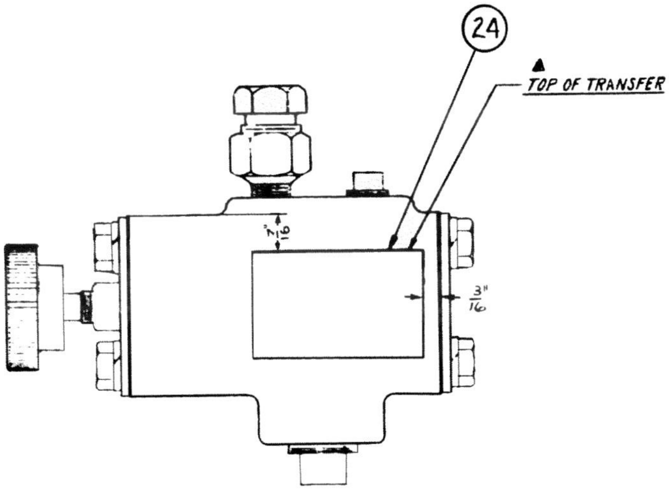

The location of the instruction decal on the transport valve on the WD and WD-45 series tractors. The top dimension is 7/16 inch from the left edge; the right dimension is 3/16 inch from the right edge of the decal to the right edge of the valve. *Drawing courtesy of AGCO Corporation*

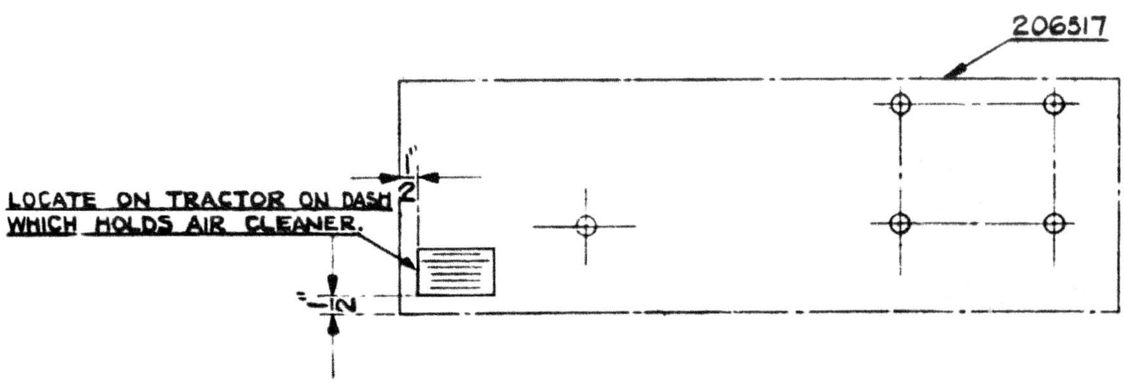

Unstreamlined WC tractors have the patent decal located 1/2 inch from the left of the dash, and 1/2 inch up from the bottom. *Drawing courtesy of AGCO Corporation*

adjusting the pump to activate certain features, such as draft control, hold position, and delayed lift. The decals were located on the right-hand side of the battery box, centered. The decals changed considerably over time, and were different between the CA and the WD/WD-45 series.

The original decal, 223670, used the words "Draft Control" instead of "Drawbar Control" on the later decals, and identified the implements that were used with each setting. The decal had a black border and lettering and a transparent background. The decal was 5 5/8 by 7 3/4 inches.

The original decal was replaced by 226171 in October 1952. The new decal included drawbar control wording, and replaced the implement usage with words describing what each setting did. Decal 227076 was introduced in July 1954, and was the same except for a Persian Orange background on the decal.

In February 1956, the WD-45 received a different decal, 228412, which added a line at the top identifying the Traction Booster system. The size of the decal changed to 6 1/4 x 7 3/4. A slightly different decal, 228411, was made for the WD. A Spanish version of the WD-45 decal, 229572, was added in early 1957.

The first CA hydraulic pump decal was 225428. The image of the pump includes the operating lever, a clear difference from the WD decals. This decal lists implements used with each pump setting. It had black lettering and border on a transparent background. The decal was 5 5/8 by 7 3/4 inches. Decal 227075 replaced the original decal, with the drawing

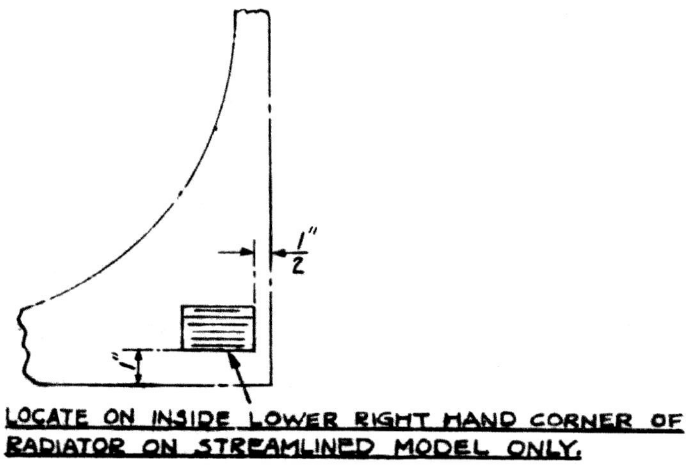

This shows the location of the patent decal on the styled WCs. The decal is 1/2 inch in from the right side, and 1 inch up from the bottom. *Drawing courtesy of AGCO Corporation*

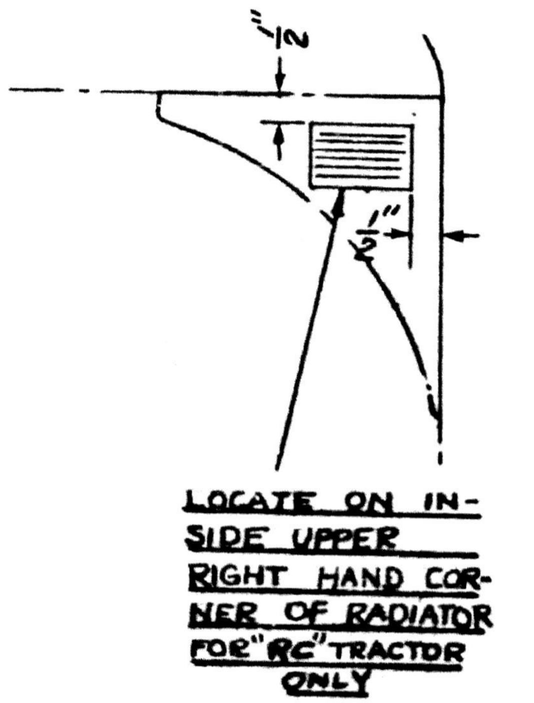

This shows the location of the patent decal on the RC. It's 1/2 inch from the top of the seam, and 1/2 inch in from the right edge. *Drawing courtesy of AGCO Corporation*

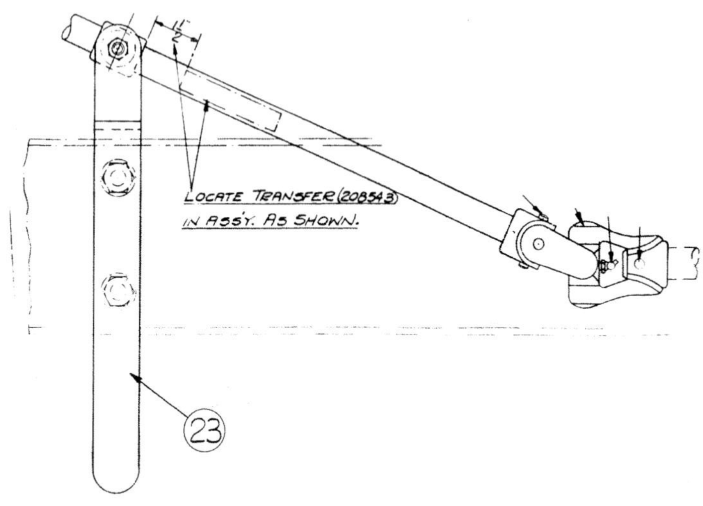

This drawing shows the location of the U-joint lubrication decal on the WC Cane tractor. WC, WD, and WD-45 were presumably similar. The decal is located 1 1/2 inches below the collar. *Drawing courtesy of AGCO Corporation*

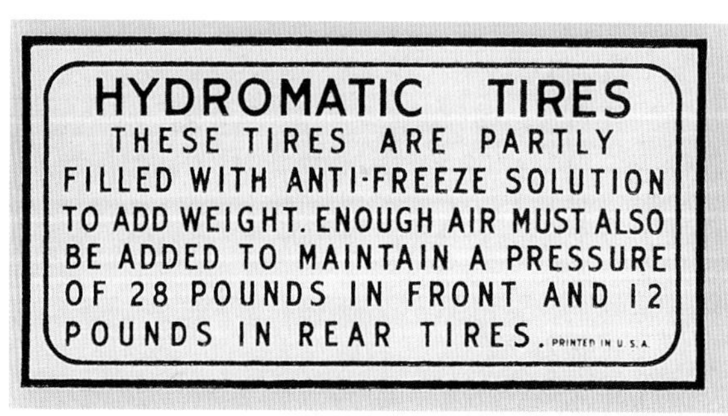

The tire instruction decal 208663-4.

dated July 1954. This eliminated the reference to specific implements, and changed the background to Persian Orange.

Two new decals, 228416 and 228417, changed the total dimensions to 6 1/4 by 7 3/4, and added the words "Traction Booster" at the top. Decal 228416 was the new solvent-applied decal for factory use, while 228417 was the water-applied decal for service use. (Solvent-applied decals were more durable, but could only be applied in the factory, not in the field, hence the water-applied service decals.) After 1960, both decals had the background color changed to Persian Orange Number 2.

Engine Designation Decals

When production switched to the WD-45 carbureted engine, a new decal was authorized to identify the new engine. The "Power Crater Engine" decal was a graphical depiction of the combustion chamber of the WD-45 engine, with a sparking spark plug and flames. The plug and "power accentuation marks" were white, the combustion area was transparent, the "Power-Crater" lettering was yellow, the "Engine" lettering was white, the piston, valves, and head were gray, the background was black, the border around the black background was transparent, with a square black liner around all. "Printed in U.S.A." appeared on the face.

A similar decal was authorized for the WD-45 Diesel engine, called the "Dyna-Swirl" decal, but apparently no tractors were produced with this decal. The drawing was canceled shortly after release.

Transport Valve Decals

The WD-45 series with the new transport valves also included a new transport valve decal, 227310. The decal specified which way to turn the adjusting knob to attain the particular hydraulic setup needed. This decal had black letters on a white background, with a black outline 1/32 inch in from the edge of the decal. The original wording of the decal said:

> Turn knob in before using pump on "Hold Position." With semi-mounted implements, use "Drawbar control" with knob out.

This decal was changed April 13, 1955 (with new decals to be used as soon as possible), to read "Turn knob out for implements using both remote ram and tractor lift arms. Otherwise, turn knob in." With the change in wording, the dimensions changed from 1 1/4 by 2 1/8 inches to 1 3/16 by 2 3/16 inches. The decal was placed on the valve itself.

Allis service decal sets did not include this decal, but the decal was included with pumps supplied as service parts.

Hydraulic Ram Decals

Hydraulic ram decals were used on Allis Bs and Cs to indicate the top of the ram. Decal 213291 had black lettering on a white background that simply said "top." There was an earlier decal, 211877. The right side of the decal was lined up with the centerline of the top pin, and it was centered top to bottom with the ram body.

Model WD Hydraulic Pump Control Hand Lever Decal

A decal reading "Keep Lever Down When Not In Use" was located on the hydraulic control quadrant.

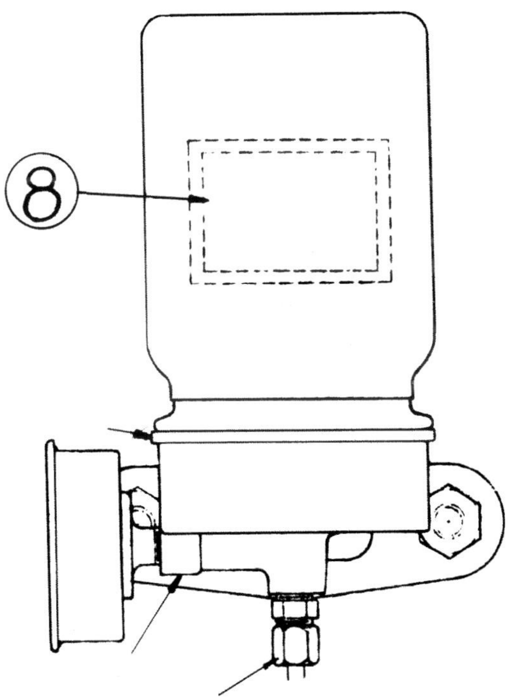

The dashed lines show the location of the oil filter decal on the glass jar oil filters used on the WC. *Drawing courtesy of AGCO Corporation*

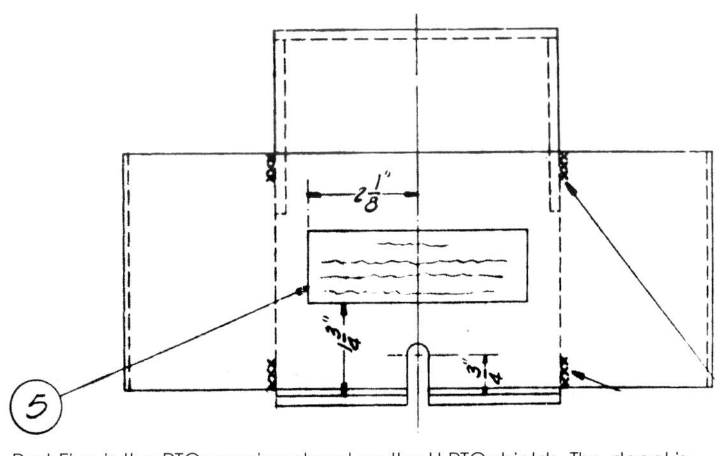

Part Five is the PTO warning decal on the U PTO shields. The decal is located 1 3/4 inches in from the end of the shield. The left edge of the decal is 2 1/8 inches to the left of the centerline. *Drawing courtesy of AGCO Corporation*

The "Snap Coupler" decal, 226684-1.

The location of the trade name decal, when the kerosene hood is used on B, C, and CA tractors. *Drawing courtesy of AGCO Corporation*

The decal had black lettering on a white background, and was surrounded by a black outline stripe, although this was located 1/16 inch in from the edge of the decal. This decal was used on the WD-45 as well.

Patent Decals

Allis did some unusual things with patent decals. Sometimes the decals were on the exterior of the tractor, and sometimes they were not. Sometimes they referred to a specific tractor, and sometimes they did not.

Decal U-3736 (later numbered 203736) was a patent decal for the UC tractor. The decal print is unclear as to when the decal was first used, but there is a change listed for April 22, 1935. On this date, the second patent number was added, 2,039,749. Previously, it said "other patents pending." The first number was 1,868,474. At this time, the general dimensions of the decal were changed. This patent had black letters on a white background, with a thin black outline with rounded corners 1/16 inch in from the edge of the decal.

Decal 206586 was put on WC and RC tractors. The original decal drawing was dated December 30, 1937. This seems to be an error, however, as three change notices are on the drawing dating before this. At first the only number on the decal was 2,073,613, but 2,098,498 was added April 15, 1937. Patent 1,641,813 was added December 6, 1937. On December 23, 1937, the prefix "U.S." was added to the patents, and the numbers were placed in numerical order.

Decal 213538 eventually was used on several tractors. The decal had blue lettering and a blue border on a white background. The decal went through many revisions as patents were added and subtracted from the decal. Patents from several nations were listed (the final version had 27 patents from five nations). This decal was used from 1942 through at least October 1956. D series patents were added

The location of the CA model decals. The rear edge of the model symbol decal is 1/4-inch forward of the seam of the grille.

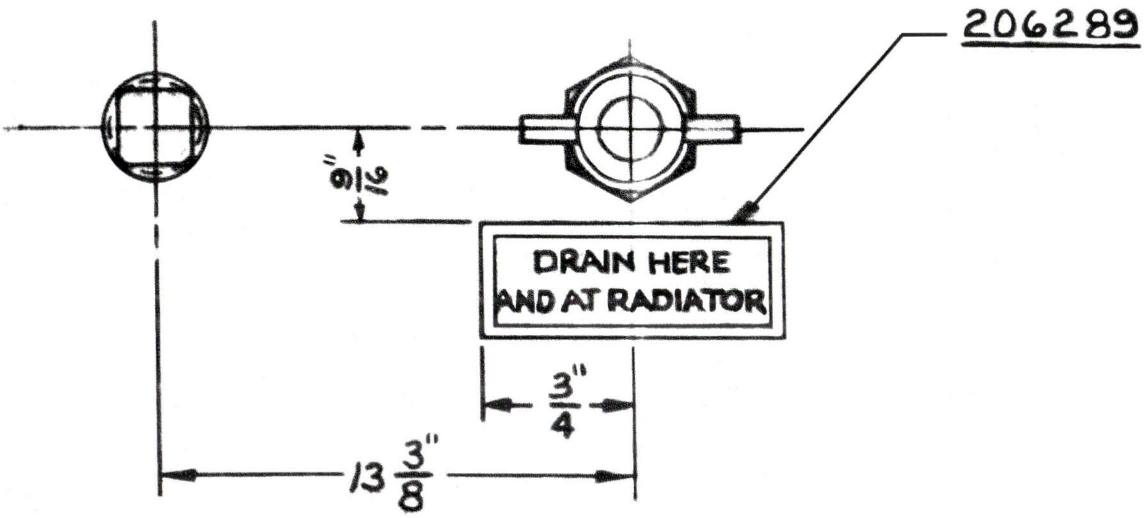

The location of the cylinder block drain decal for the WD-45 Diesel tractor. *Drawing courtesy of AGCO Corporation*

sometime after February 1957. Patents added for the D series include numbers 2,777,375 and 2,779,604. Some WD-45, B, CA, and IB tractors may have this patent, as it was added after the then-current stock had expired, and the earlier tractors may still have been in production. Patent numbers 2,834,277 and design patent 182,147 were added after August 1958, and decals with these numbers should be considered too new for tractors discussed in this book.

Location of the patent decals varied by tractor. IB tractors had the decal located on the left side of the fuel tank shroud, with the bottom edge located approximately 5 inches above the bottom of the shroud, and the right edge of the decal lined up with the left corner of the square plate on the shroud.

B, C, and CA tractors had the patent decal located on the right side of the shroud, 3 1/4 inches above the bottom of the shroud, and 2 1/2 inches to the rear of the right edge of the middle hole of the three small holes on that side of the shroud.

WD and WD-45 tractors had the patent decal located on the right side of the fuel tank support, with the bottom edge of the decal located 3/4 inch above the bottom of the support, and the rear side of the decal located 3/4 inch in front of the front edge of the larger lightening hole.

Steering Wheel Universal Joint Decals

The WC family of tractors had a "Grease Universal Joint Regularly" decal that was drawn in July 1938. This decal was used in WC, WD, and WD-45 up to the end of production. Decal 208543 had black lettering on a white background. "Printed in the USA" was added to the decal in September 1939. The decal was placed so it could read correctly from the right side of the tractor, on the steering shaft, 1 1/2 inches down from the lower steering shaft support.

Hydromatic Tires Instruction Decals

Decal 208663 was the instruction decal that listed tire pressures for liquid-filled tires. The decal has black letters on a white background. "Printed in the USA" was added to the decal drawing on October 17, 1950.

Decal 227496 was nearly identical, except that the rear tire pressure was increased from 12 pounds to 14 pounds, and the lettering was arranged slightly differently. This decal drawing was completed in late August 1955.

In general, one decal was used on the front of the tractor and one on the rear, although both wheels may have occasionally had a decal. Which side seems not to have been recorded. On a front wheel, when the valve is at the top of the wheel, the decal goes on the flat disc directly underneath. On the rear tires, position the valve on the bottom of the tire, then place the decal under the valve on the rim itself. These positions may not be universal.

PTO Guard Decals

PTO guard decals all seem to have been 210300, first drawn in February 1939. The decals were added in response to a Farm Equipment Institute initiative. The later (post-1952) decals had black lettering on white backgrounds, surrounded by a 1/16-inch black stripe. The left edge of the decal is located 2 1/8 inches from the centerline of the shield (on the square sheet-metal shields with braces). The decal is located 1 3/4 inches in from the inside of the lip of the shield. These shields were used on the B (released in February 1940), C, IB, and CA tractors.

Engine Block Drain Decals

The original engine block drain decal for the WC tractors was drawn in March 1937. Decal 206289 had text reading "Drain Here and at Radiator" with black lettering on a white background. There was a border on the decal. The "Printed in USA" was added in mid-November 1950. This decal was used up until the end of WD-45 production. The decal was placed with its center aligned with the center of the drain cock, slightly underneath the drain cock. When looking straight on toward the engine, no part of the decal should be obscured by the cock.

Oil Filter Decals

Decal U-4003 (later 204003) was drawn in late October 1935, and was the first oil filter instruction decal for the W engine. The label read "Replace Filter when crankcase oil darkens. Ordinary life of filter with gasolene 200-400 Hr. With Kerosene 100-200 Hr. -Allis-Chalmers." The decal used black lettering, a white background, and had a thin black outline 1/8 inch from the edges. The decal was originally used on the glass oil filter jar, where it was positioned to leave an equal amount of space between the top and bottom edge of the cylindrical wall of the jar. The decal was paper gummed on the printed side. This decal was superseded by 204861 in October 1936.

Decal 204553 was originally drafted in November of 1932. This decal was for Purolator oil filter instructions. The decal used was supplied by Purolator, and had a Persian Orange background.

Decal 224973 was drawn June 20, 1950. The final version of the decal, placed on the oil filter can, read "Change Filter every 100 Hours, Allis-Chalmers, U.S. Patent No. 2287982." "Printed in U.S.A." was added October 17, 1950. The height of the decal was reduced to 1 1/8 inches from 1 7/16 inches on April 18, 1952. The same date, the decal was widened and the instruction data was changed to the final text. On January 29, 1957, patent numbers 2078525 and 2179784 were removed. This decal had black lettering on a transparent background with a black border located 1/16 inch in from the edge of the decal.

Air Cleaner Instruction Decals

Air cleaner decals were provided by the manufacturer of the air cleaner. Allis did not always consider the decal to be a separate part, but on earlier tractors that used air cleaner decals, Allis included the decal as part of the assembly. The air cleaner decals could change completely and still be carried under the same Allis parts number.

Decal 209263 "Service Daily" was used on Donaldson air cleaners on B, C, and CA tractors, if not others, starting in 1939. The Donaldson part number was P-2745-7, at least in 1952. The drawing for the decal was completely redrawn at least once. The shape was rectangular with rounded corners, and featured the Donaldson logo at the top center of the decal. The decal gave the oil weights to use in the oil cups for various temperatures. The decal was 1 13/32 inches tall, and 2 15/16 inches wide.

Also seen on B, C, and possibly CA, if not larger tractors from 1939 through at least 1952, was an "Oil Level" decal, which had the Donaldson logo between the two words. The part number was 209264, and the decal was 2 1/2 inches wide by 3/8 inch thick. Two curved arrows at each end pointed up.

Decal 226629 was the later air cleaner service instruction decal. One version was made by United Specialties, while the other version was made by the manufacturer of the other air filter, probably Donaldson.

Wiring Diagram Decals

The WC tractors used wiring diagram decal 208401. This decal was 2 5/8 by 4 inches in its outside dimensions, with black lines on a white background, and was originally drawn in June 1938. "Printed in U.S.A." was added in the lower right-hand corner of the decal in February 1939. The decal lived into the WD and WD-45 series tractors.

Allis Cs and Bs used wiring decal 212227. This decal was 2 5/8 by 4 inches, and had black letters on a white background. This decal was originally drawn in December 1940. In February 1951, "Printed in U.S.A." was added in the lower right-hand corner.

Allis Gs used decal part number 800106. The decal had "Allis-Chalmers MFG. Co., Gadsden, Ala." in the lower left-hand corner, and a complete view of the wiring system. The decal was 2 5/8 by 4 inches wide, with black lines on a white background.

The wiring diagram decal was located on the inside of the battery box top cover.

Battery Filling Decal

Decal 210090 was an instruction decal having the text "Fill battery to not more than 1/4 inch above plates. Use distilled water for best results. Clean rain water next best." The lettering was black on a white background, with a thin black outline 1/16 inch in from the edge of the decal. "Printed in U.S.A." was added in mid-October 1950. This decal was located on the outside surface of the battery box lid. The original drawing was dated August 25, 1939, and was used all the way through W series production. The battery instruction decal was located on the outside of the top cover of the battery box.

Snap Coupler Decals

Model B, CA, WD, and WD-45 tractors with Snap Coupler hitches had an instructional decal. The decal had a black background with yellow lettering across the top saying "Snap Coupler," below which was a drawing of the bottom Snap Coupler hitch receiver. The chain and bell assembly were transparent, while the accent marks showing the attaching point were white. The decal was located on the left side of the top of the bellhousing running lengthwise with the tractor. The bottom edge of the decal rested on the fillet edge of the bellhousing, while the upper right-hand corner of the decal was located 1/2 inch from the open end edge of the bellhousing.

Torque Tube Cover Decals

Models B, C, and CA tractors had a decal saying "Remove cover to grease clutch shifter." The decal had black letters on a white background. The decal was 3/4 inch high by 1 13/16 wide, and had a thin black border 1/16 inch in from the edges. The decal was first drawn in July 1938, with the words "Printed in the USA" added, probably in 1951. The decal appeared on the round torque tube cover, and was located halfway between the center and the edge of the round cover.

OIL PRESSURE
ALLIS-CHALMERS
TRACTION BOOSTER
ALLIS-CHALMERS
WATER TEMPERATURE
220
ALLIS-CHALMERS

CHAPTER 3

General Parts

Allis-Chalmers used certain parts on several different tractors. The parts tended to change over time, including specifications and suppliers, but were kept under one parts number for ease of record keeping and simplification of inventory.

Electrical Systems

For the most part, Allis used asphalt-impregnated loom during the period in question. Spark plug cable retainers (cable ties) were made of rubber, had a 5/8-inch internal diameter, a 1 1/4-inch outside diameter, and were 1/4 inch thick.

Gauges

Ammeters

Allis tended to use standard ammeters across the product lines. The ammeters were, for the later tractors, carried under part number 208302. However, this number was actually a specification, rather than an actual part, and it included ammeters from several manufacturers, as well as clamps and hardware. (Allis liked having multiple suppliers for both assurance of supply, and to have them bid against each other to keep costs low.) The ammeters changed over time.

Engineering departments and corporate accountants like to produce and stock as few parts as possible. Parts that do the same thing tend to get used on many different tractors, instead of just one particular tractor. The gauge cluster on the WD-45 Diesel shows the temperature, Traction Booster, and oil pressure gauges. All of these gauges were used on many different models of Allis-Chalmers tractors.

The last two ammeters used under 208302 were Allis part number 226604 (Rochester Mfg. Co. 4015-C15) and 226603 (Stewart Warner 357-J). These ammeters were incorporated into the specification in December 1954. The 226603 used a black lithographed background with white markings, a white pointer, and a glass lens. The gauge said "Amperes" under a -20 to +20 scale. A Persian Orange bezel was specified. On the scale, tick marks appeared every five amps, but no numbers were used except the 20s. No "Allis-Chalmers" appeared on the face. The pointer had a small diamond-shaped section in the middle.

The 226604 used a stainless-steel bezel, with the words "Discharge" and "Charge" under the scale. Tick marks covered every two amps, with 10-amp markings located between the 20s. There was no diamond shape on the pointer.

Oil Pressure Gauges

The 207834 oil gauge was used throughout Allis tractor production from the late 1930s on. This gauge had a range that covered 0–30 pounds, but was not marked in increments. Instead, a short range on the left covered 0–5 pounds, then another range ran from 5 to 30. In the arched rectangle was the word "NORMAL." The words on the dial were bold black on a white background. The pointer had a black finish—although another color may have been specified in early production—and pivoted from the bottom of the dial. The words "Oil Pressure" appeared in two lines over the pressure range, with "Allis-Chalmers" appearing under the range in a straight line just under the centerline of the gauge. Again, some changes may have occurred in early production. The dial changed in 1948 and 1951.

Although discussed in other chapters, headlights, radiator caps, fuel caps, and mufflers tended to be used across several different models, although not as many as the water temp gauge. Decals were also shared among many similarly sized tractors.

The bezel for this gauge changed several times. The original bezel had a brass finish, but this was changed to black in January 1940. The bezel was then changed to Persian Orange in September 1949, and was again changed to stainless steel in May 1957.

The 206959 oil gauge face had bold black characters on a white background. The pointer had a black finish. The bezel was originally a black finish, but was changed to Persian Orange in September 1953. The bezel color was again changed to stainless steel in December 1953. The gauge was manufactured by Rochester Mfg. Co. after 1948. Manufacturer of this gauge was not specified before 1948, indicating that there may have been more than one supplier. The gauge had "Oil Pressure" across the top, with a 30-pound range, with ticks and numbers for every 10 pounds. "Allis-Chalmers" was written across the bottom.

On Allis Us, tractors before U-15344 had the oil gauges mounted on the operator's panel. After U-15344, the oil gauges were mounted directly on the oil filter. On the tractors with the gauge mounted on the filter, different gauges were used, depending on whether the tractor had electric lighting and starting.

Oil gauge 200197 (original number U-197) first called for an Allis-Chalmers marking, but the specification was changed to United on March 13, 1929. On April 4, 1929, it was changed back to Allis, leading to confusion over how long the United name was put on the oil gauges, if it ever was. The gauge was marked in 25-pound increments up to 75 pounds. This gauge was used on all U and UC models with Continental engines, and on Us and UCs produced before September 1935. It may have been used on some other tractors, but was discontinued on tractor U-15344. These gauges were made by U.S. Gauge Co., or Rochester.

Oil gauge 203805 (original number U-3805) was first drawn on September 9, 1935, and was graduated in 10-pound increments up to 30 pounds. This gauge was also used on Us and UCs with the Allis engine, but was discontinued before tractor U-15344 as well. This gauge was made by Rochester.

Oil gauge E-709 was used on the 25-40. Also possibly used on the Es was gauge AM 8728, later renumbered 238722. This was a 75-pound gauge, with the graduations running every 25 pounds. "Oil Pressure" was located under the scale, with "Allis-Chalmers" on the face on two lines below that. These gauges were provided by Rochester Diaphragm Gauge or U.S. Gauge Co.

Tachometers

Tachometers were carried as toolbox items, not attached to the tractor. Us and Es had tachs made by Sun Electric Corp. The tach measured from 0 to 2,000 revolutions per minute, and was designed for a four-cycle magneto ignition. The Sun part number was J-12, which was carried by Allis under part 222301. The

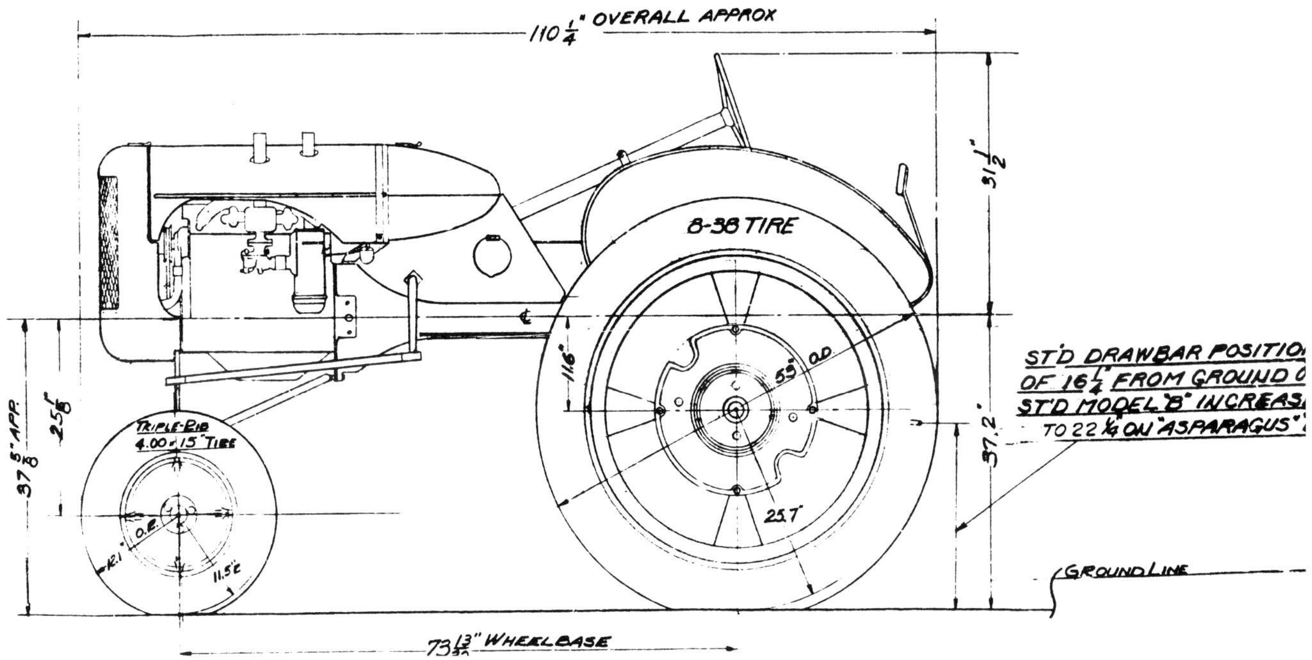

The B Asparagus was a high-clearance version of the B. It offered 6 more inches of ground clearance under the rear drawbar than the standard B. This drawing dates from August 1939. *Drawing courtesy of AGCO Corporation*

Allis print doesn't show the face of the tach, so it may have had Sun's standard face. These tachs may have been used mainly on power units.

B and W engines also had a tach available, a Sun model J-5, carried under Allis part number 222300. This tach had a scale of 0–3,500 rpm, again for use with a magneto ignition.

Traction Booster Gauge

The traction booster gauge allowed the operator to monitor and adjust the hydraulics to change the weight transfer between the implement and the tractor, thus gaining traction when needed. The operator lowered the depth of the implement to the desired working depth, then raised the hydraulic lever slowly until the gauge barely registered. If hard working was encountered, the lever was raised a little more, then placed to the original location when the load passed. The gauge assemblies were painted orange, with a Persian Orange gauge bezel, with a white gauge face with black lettering. The pointer was black. The first gauges had a range of 30 degrees on either side of vertical, with gauges made after February 1954 going to a 33-degree range. The words "Traction Booster" were arched and curved over the top, with a range of small boxes underneath. "Allis-Chalmers" ran in a straight line on the bottom of the gauge.

Water Temperature Gauge

The part number listed for most, but not all, Allis water temperature gauges in the final parts manuals was 238548. There were two different designs (having their own, older parts numbers that did not appear in

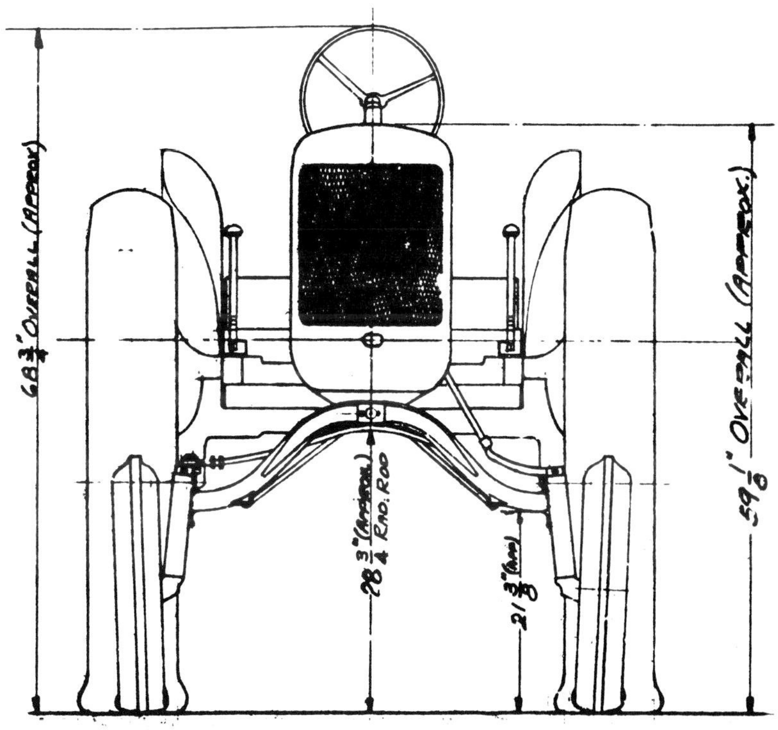

The front spindles and tires appear to be much larger than the standard B. The B in this drawing still has the hand brakes. *Drawing courtesy of AGCO Corporation*

parts manuals) and three different manufacturers for this gauge, as well as at least two different parts numbers. The main design was originally a Stewart Warner gauge that was also supplied by the AC Spark Plug Co. The optional gauge, shaped like a horn mouthpiece, was supplied by Rochester, later known as American Standard Controls Division.

Lift Arm Latches

The Snap Coupler system used two lift arm latches on the hydraulic lift arms to grasp the arms of an implement. Allis-Chalmers revised these arms several times in production. The first style of latch, used before March 1954, was made in two halves and bent together. The second latch assembly, used from March 1954 through approximately June 1956, used a one-piece latch, with the main body having a plain, nonreinforced hole where it attached to the hydraulic arm. The third type of latch assembly, July 1956 to May 1958, used the same latch, but with a reinforced attaching hole. The fourth type of assembly, used in May 1958 and after, had a large additional hole underneath the regular attachment hole.

Radiator Curtains

The original radiator curtains used on Allis tractors were made of 8-ounce olive drab waterproof cotton duck.

Seats

Allis used a variety of seats over the years, but many of these went on more than one model.

Seats for Models E, U, IU UC, WC, RC, WF, IB, W-SP

Originally, at least a few early WCs used an upholstered seat. The seat, 203591, was made by the Milwaukee Saddlery Company, and was upholstered in two options: cowhide and horsehide, to see which would work best in the field.

Apparently neither was the answer, as further production used a steel pan seat, made in at least two different styles by two different manufacturers. The 200145 was originally titled U-145, and was used by most of the large Allis tractors. The pan was used on E, U before tractor U-12000, the UC before tractor UC-2819, WC, WF, and IB models. The general dimensions were held constant to use the same seat pad, 204229. The seat pad had two optional designs, which themselves had optional materials. The seat pad was normally covered in Number 6028 DuPont Cavalon, but after April 1942, if that material was unavailable, green or olive drab 12-ounce waterproof duck could be used. The seat was trimmed in orange fabricoid cord or, after May 1943, in green 8-ounce waterproof duck. The retainer was originally the DuPont Cavalon, but this could be replaced by the 12-ounce duck, or a 70-pound-test drawstring could be used after 1942. Seat pads were mandatory with steel rear wheels before 1942.

Later Us, UCs, most IUs, and all As used seat frame U-3550 and seat cushion U-1351. The seat frame was pressed steel and included a back and side arm rests. Several suppliers provided the U-1351 seat cushions. The usual color after December 1945 seems to have been olive drab, in waterproof duck material. Earlier cushions were available in black pontine fabric with orange trim cord, and another version, for which specifications could not be found. Some early IUs used seat cushion U-574, which was covered with a "good grade of automobile cloth of Spanish Brown."

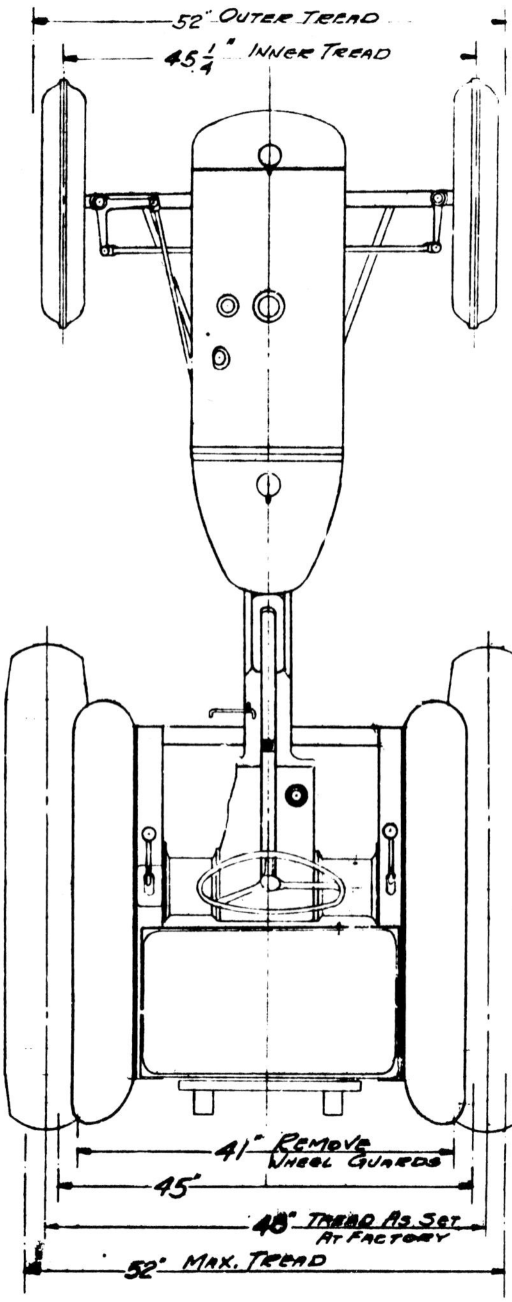

B Asparagus tractors could have the rear tread narrowed to as little as 41 inches if the rear fenders were taken off. The front tread, however, was fixed. *Drawing courtesy of AGCO Corporation*

Some Us built in 1936 and after may have come with seat cushion 204472. This cushion had black Number 70 Pontine Cushion Top covering with orange fabricoid cord.

On the IU, a seat back pad was also used. The first seat back pad was U-575, but this cushion was replaced at some unknown point. The second seat back pad was U-1357, but it was later replaced in May 1936 by 204910. Seat back pad U-575, like seat cushion U-574, was covered with a good grade of automobile cloth of Spanish Brown. Back pad U-1357 was made of imitation black leather in Spanish Black with orange cord. This was fastened in place with grommets. Back pad 204910 was covered with black Number 70 Pontine Cushion Top material with orange fabricoid cord, and was held in place with two black leather straps with buckles. The straps were riveted to the cushion.

Seats for Models B, C, IB

The B and C seat was an internally sprung bench seat. The IB seat was an externally sprung pan with cushion. The IB used seat cushions and backs identical to the W family.

B seat frames differed by width, with the regular tractor, the 36-inch tread tractors, and the asparagus tractors all having different seat frames and seats. Cs before C-109 had a different frame assembly than later tractors, which used the seat assembly from B production tractors B-12016–B-52718.

The main seat cushion for the B (205693) was changed many times in production, although the part number didn't change because interchangeability was not affected. The original seat material was black Number 70 Pontine Cushion Top, but a decision dated August 26, 1937, changed the material to green. The material was later described as 12-ounce olive drab duck. The internal spring types were changed several times. The seat was originally provided by both Milesco and Bostrom, but Milesco was later dropped. An optional construction adopted in July 1949 offered polyfoam filler, using a block vinyl cover with a double seam, and covered with a clear view polybag. The part number was stenciled on the cushion base.

The first 100 Bs used a hardwood seat back (205695), finished all over. Tractors B-101-B-12016 used a cushioned seat back 207354, covered with 12-ounce green canvas cover, with color matching the seat cushion. A formed sheet-metal plate was folded around the back of the seat. After B-12016, a pressed steel back was used. The pressed steel seat back was also used by all Cs.

Seats for Models WD, WD-45

The WD and WD-45 both used a sheet-metal seat. The shock absorber was provided to Allis in black finish by Monroe. Available as a special option was detachable seat pad 226173. This seat was made of 24-ounce white cotton belting duck, coated with black Koroseal.

Seat for Model G

The G used the standard pressed steel seat, 225050, with a simple leaf spring underneath.

Seat for Model CA

The CA used a sheet-metal seat. The original seat was 218102, but this was superseded by seat 225050. The shock absorber was the same as the WD and WD-45 absorber and was provided in black by Monroe.

Umbrellas

While a variety of umbrellas were probably available, one common design is green and orange panels alternating with the dealer imprint on them. The green was probably olive drab similar in color to other green fabrics used in seats and umbrellas.

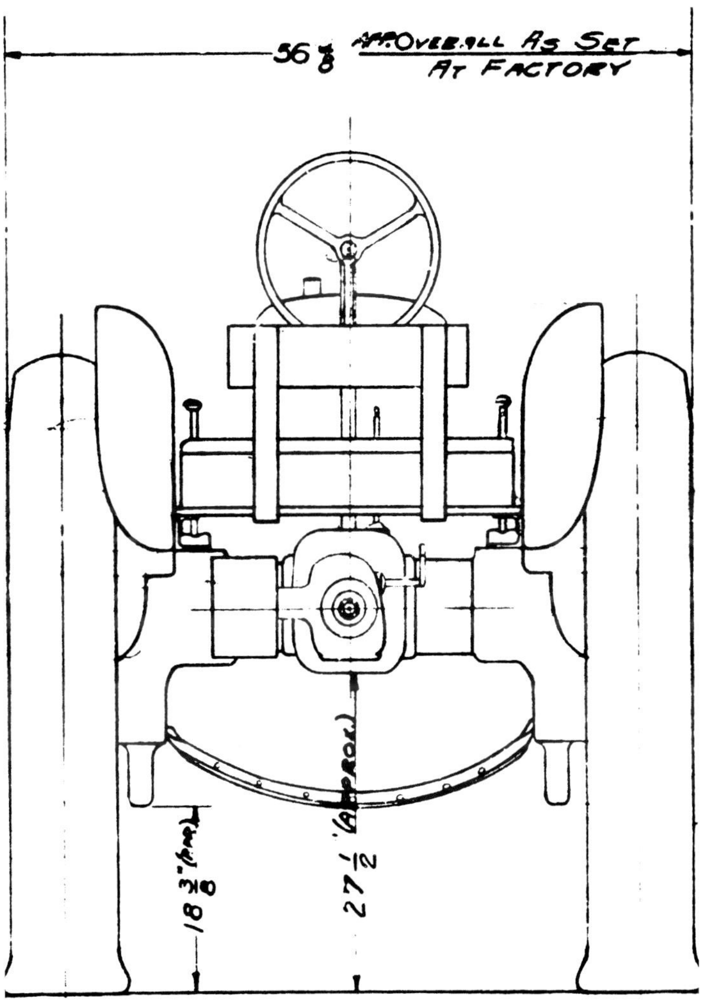

The rear view of the B Asparagus shows the crop clearances available. Asparagus is a tall, high-value crop. Growers of this specialty crop could afford the premium price of a special tractor. *Drawing courtesy of AGCO Corporation*

The Model E in all its permutations was the tractor that put Allis-Chalmers firmly in the tractor business. The ultimate development in the Model E, which over the years included many different developments, was the 25-40 Thresherman's Special. Fitted with a special engine so powerful that the company refused to warranty the tractor for plowing purposes, the Thresherman's Special was the production hot rod of Allis tractors.

CHAPTER 4

Model E 25-40 and Model A

The Model E was Allis' first commercially successful tractor. Earlier tractors had suffered from unsuccessful design or high prices, or both. By the mid-1920s, Otto Falk, Allis' president and the man responsible for Allis-Chalmers' move into tractors, decided to make a change, and brought in Harry Merrit. Merrit was an experienced tractor man who was told to make the tractor division successful or close it out. By reducing the price of Allis' main tractor, Merrit discovered that Allis could sell enough tractors to make the business worthwhile.

But could Allis produce a tractor cheap enough to sell at the new low prices and still make a profit? To find out, Merrit had the tractor redesigned. The first redesign was not done by an engineer, but rather the tractor advertising manager. The chief engineer, Max Patitz, understandably blew his fuse at this occurrence, and demanded the right to come up with his own redesign. The two tractors were then stood side by side, and the best features of each were incorporated into the new tractor.

The result was that the E, which had been in production since 1918, first as the 18-30 and then the 20-35, was reborn as a lighter, cheaper tractor. The new 20-35 was really the company's first commercially successful tractor.

In 1930 a new improved version of the E began production at Serial Number 24186. The 25-40 horsepower Model E had a 5 x 6 1/2 bore and stroke, as opposed to the 20-35's 4 3/4 x 6 1/2. The 25-40 remained in production until 1936, when it was replaced by the A. The engine of the 25-40 E remained in production as a power unit for many years afterward.

The As were created by combining the 25-40 E engine with the cheaper U transmission and platform. However, the U transmission seems not to have been able to handle all the horsepower, so at Serial Number 26526 a new, heavier-duty transmission was used.

Variations—A Industrial and AO

Only a few A industrials were built. These tractors, according to the release notice, differed mainly in tire tread, although there were at least a few other differences. Some tractors had individual brakes, and some may have had foot accelerators.

Allis seems to have built at least a few AOs, starting in late December 1938. This tractor had the Hesselman fuel-injected, spark-ignition engines from the KO crawlers. Allis built several models of crawlers with these engines in the late 1930s, but finally abandoned them in favor of Detroit Diesels. Photographs exist of AOs, as well as the original Allis change notice authorizing the tractor. The engine was different, as was some of the sheet metal. Most, if not all, of these tractors were exported. An AO is being re-created by a collector.

Engines

The E 25-40 and the A shared the same basic engine. Two different intake manifolds were available for the 25-40—the regular manifold and a special manifold for a "special type heated carburetor." The special manifold also had its own exhaust manifold cover. The As received their own modernized manifold. Carburetors differed between the tractors, with 25-40s using carburetor E-1657. The special heated carburetor was E-1935 (probably a Kingston Model M 1 1/2 inch kerosene special), while the A used 213019. The As were built in both gasoline and a kerosene versions.

E 25-40 tractors were also built in what was called a "Thresherman's Special" with a 5 1/4 x 6 1/2-inch

The 25-40 used a post with spring type of steering support. Although the Thresherman's Special was not guaranteed for plowing, it was fitted with the standard 25-40 drawbar, and was actually quite good at plowing. The long lever on the left side of the platform is the clutch lever. The E didn't have a transfer on the rear of the fuel tank, and may not have originally had them on the rear of the fenders either.

With its low front axle and larger horsepower, the Thresherman's Special was clearly not intended for row crop work. Built mainly to pull threshers from job to job and to provide the belt pulley power when they got there, the Thresherman's Special was the last of the breed, a throwback to the earliest days of steam traction engines, when plowing with power was but a dream.

The 25-40 had some decent-sized front wheels for a tractor of its time. The starting fuel tank seen on the left fender was relocated there by a previous owner to improve the fuel flow. A Purolator decal is on the oil filter.

The Es used a base mount magneto, from several different suppliers. The orange piece under the magneto is the base on which it was mounted. On the original tractor, the carburetor was probably painted, although no Allis paint specifications have been found that early.

A top view of the E, showing fuel cap (painted orange) and the radiator cap. The radiator cap used a wire bail to hold it down. The same caps were used on the A. This E was used every spring for many years for plowing a few acres at the Rock River Thresheree. The owner delighted in getting there before everyone else (which meant it was still dark) and starting. Reportedly, flames were shooting 6 feet high out of the exhaust stack the morning this photograph was shot. No wonder there are a few scratches on the paint!

A choke rod is run forward so the operator can work it while cranking the tractor. The large size and compression of the Thresherman's Special engine means a tough crank, and anything to make the engine start the first time is necessary. The author tried to crank the tractor, and got it around twice before giving up. Here, the expert takes over. Note the base mount magneto.

stroke engine, which was essentially the same engine as the regular 25-40 but with different sleeves and pistons. These tractors were intended for belt pulley work, and the warranty did not cover plowing. They are fairly rare; one estimate has it that about 70 were built. A few collectors have plowed with them and the results are spectacular—plenty of power, and very impressive exhausts.

Engine Serial Number Suffix Codes

Allis provided engines with various compression ratios to ensure that full engine power was available to all users, including those who farmed at elevations where low air densities adversely affected horsepower.

A Engine Codes

(Engine serial numbers are located on the left side of the block)

4 3/4 x 6 1/2 engine

G= 4.68:1 compression ratio

5 x 6 1/2 engine

K=4:1 compression ratio (kerosene under 3,000 feet altitude)

G=4.45:1 compression ratio (gasoline under 3,000 feet altitude, kerosene 3,000–6,000 feet altitude)

A=5.2:1 compression ratio (gasoline 3,000–6,000 feet altitude, kerosene 6,000 feet altitude and higher)

B=5.8:1 compression ratio (gasoline over 6,000 feet)

Fuel Tank

The 25-40 used auxiliary fuel tank E-1863. The fuel line leading out of the tank to the fuel filter was 3/16 inch on the E versus 1/4 inch on the A. The E used fuel cap U-204 while the A used 204827. The A used 1-gallon fuel tanks below Serial Number 26526 and 1 1/2-gallon fuel tanks after that serial number.

In general, Allis painted everything orange that was on the tractor at that point on the assembly line, as shown on this 25-40. It is difficult to determine at this late date what was or wasn't on the assembly line at that point, and what was masked off and left the color of the part as supplied to Allis.

Parts books drawings do not show diamonds on the rear of the fuel tank of the 25-40, where they do show them for the A. Sales literature and photographs do show decal locations in some instances, but since decals changed in both their location and appearance, it's hard to tell what was original for these very early tractors. The E had an upright post with spring seat support, with a pan seat.

The operator's platform on the 25-40 was dead simple. The clutch lever is on the left side, with a brake pedal. The transmission filler plug is on the transmission, with three control levers on top. The right one is the throttle.

Fuel Filter

Sediment bulbs were available in both metal and glass.

Air Cleaner

The 25-40 air cleaner assembly had a long stack next to a round cleaner assembly. Air passed down the stack to a base, and then up into the cleaner itself.

The A air cleaner had a stack that entered at the middle of the cleaner canister. The cleaner assembly was mounted on a large cast bracket that also functioned as the air passage to the engine.

Engine Control

On the 25-40, the governor control rod had teeth on the bottom of a flat rod that extended through the dash. The teeth engaged the dash, while a spring-loaded pin kept the rod pushed down on the dash. To change speed, the operator lifted up on the lever, then moved the lever in or out.

On the A, a quadrant-with-lever system was used, identical to the later U, WC, and early B tractors.

The last of the large, standard-type tractors for Allis was the A. Using an engine derived from the E, with a transmission and other parts from the U, the A was intended for wheatland operations and other types of farming needing a standard tractor bigger than the WF.

Chassis
Model E, from the Advertising Literature

Overall Dimensions—Weight 6,000 pounds, wheelbase 90 1/2 inches, normal engine 4 3/4 bore 6 1/2 stroke

Transmission

The shift knob on the 25-40 was integral to the shift lever, while a separate knob was used on the A, which essentially used the same system of shifting as the Model U. A dust boot was used on the A where the shifter lever entered the transmission top cover.

Extensive transmission changes were made at A Serial Number 26526. On tractors before this number, third gear is to the left and back, while on tractors with this number and later, third is to the right and forward. First on the low numbers is center and back, while on the higher number tractors, it is center and forward. Second and reverse are different, while fourth was unchanged.

Transmission shafting was changed at this point, as well as the adjustment methods. Gearing was also different.

Built in an era when most demand was for row crop tractors, the A with its low crop clearance was built in very small numbers—only 1,225 were built, according to Swinford's *Allis-Chalmers Farm Equipment*. Now, As have become quite popular with collectors, due to their uniqueness and scarcity.

Tractor Speeds Before Serial 26526 (probably with 13.5x28 rear tires, although the Allis chart from which this was taken did not specify)

First Gear	2.6 mph
Second Gear	3.5 mph
Third Gear	4.5 mph
Fourth Gear	9.5 mph
Reverse	3 mph

Tractor Speeds Serial 26526 and after

First Gear	2.5 mph
Second Gear	3.75 mph
Third Gear	5 mph
Fourth Gear	9 mph
Reverse	3 mph

Clutch

The 25-40 used a hand clutch lever arrangement, while the A used a clutch pedal. The A's clutch pedal assembly had a curved pedal with a footpad that was adjustable with a setscrew.

Unlike the E, the A had an Allis-Chalmers decal on the side of the hood. The As did not have a base mount magneto like the Es, but a flange mount magneto like this one.

This particular Model A had been involved in a serious accident, so a few of the parts seen here were from a crawler tractor. Allis used similar engines in its crawler tractor lines, produced in Springfield, Illinois. The oil filter on the As had the "wing" type handle on the top, while the Es did not.

The Model A gear shift knob shows the shape of many Allis knobs, kind of a flattened shape with ridges around the middle. The knob was probably painted orange from the factory. The diamond is in the middle of the fuel tank end, as it was originally, and an original Allis gauge is at the lower right.

Model As came with two different types of lighting systems. This headlight is from the "old" system, in which can-type lights were used. (Newer As used teardrop-shaped light housings.) The "A" on this radiator is something the owner cooked up. Model As seen today are never found with the "A" on the radiator—the diamond is there, but without the letter on the inside. One piece of sales literature, however, does show the "A" on the tractors.

The thin horizontal tube is the Model A's intake manifold, with the exhaust manifold above it, hidden by the hood. Front wheels on the A seem to have been always black.

The seat on the A was part U-3550, found on many Allis tractors. The seat had a back, unlike many contemporary tractors, and probably added quite a bit of comfort. The original tire tread on the tractor probably would have had lugs at 45-degree angles.

A 12-inch single plate clutch was used on As before Serial Number 26526. On tractors with Serial Number 26526 and after , a 14-inch clutch was used.

Hoods

The 25-40 used one hood independent of fuel type. The A used a hood for low-octane fuels with holes for the auxiliary fuel tank, and a gasoline hood without the extra holes.

Electrical Systems

The 25-40s used the base-mounted Eisemann GL-4 magneto. Fairbanks Morse magnetos were sold by Allis dealers for replacements. Spark plug wires were carried in a cable tube. The As used a flange-mounted Fairbanks Morse magneto. A spark plug cable tie was used near the magneto cap, while a cable support was used further up. The As also used a spark plug cable tube identical to the 25-40s.

The 25-40 was available with a Bosch batteryless lighting system that featured two headlights and a tail-light. The Bosch system had the switch located on the generator itself. Also available was a generator-battery lighting system in either 6 volts or 12 volts. This system had two headlights and a taillight, and it also had a starter. The first As had the same system (known as the "first type") in 6-volt only. All of these systems had the large cylinder-type headlight housings. This equipment was also available on the U family of tractors.

The "second type" of lighting on the As was 6-volt and included teardrop-shaped headlight housings, a switchbox with an ammeter, and a taillight.

E battery boxes were made from wood, sandwiching the battery in, rather than being a fully enclosed box. The As had similar battery boxes. The wood seems to be oak, although the original Allis parts drawings were not found.

Steering Gear

The 25-40s and the As used a drag link steering system with the actual gearing being located just under the steering wheel on the operator's platform. The agricultural As used a longer cam and shaft assembly in the steering than the industrial As (36 1/32 inches versus 26 25/32). The tube the worm gear rode in was also different. The steering wheel on both the 25-40 and the A was part number U-141, a common Allis unit shared with many other tractors.

Brakes

Band-type brakes were used on these tractors. Foot brakes were used on the 25-40, while a hand brake was used on the A. A ratchet system using a lever engaging a slot in the dash was used as a brake lock on the 25-40, while a toggle on the hand lever was used on the A.

PTOs

The 25-40 PTO was enclosed in a long cast-iron housing. At the end of the housing, a short lever shifted the PTO into and out of gear. A cap protected the end of the shaft when not in use.

On the A, a much shorter housing was used. The PTO shifter was a short rod that connected with a longer wire rod with a loop in the end. A sheet-metal PTO guard with a cast bracket was used until approximately the start of 1940, when Allis began using an all-sheet-metal guard with braces (and warning decal) that went over the top of the end of the shaft.

Belt Pulleys

The E used a part number E-1600 belt pulley, which had a 13-inch diameter and 8 1/2-inch face, and was made from cast iron.

On the A, an offset-type belt pulley housing was used to gain some clearance over the front axle. A short belt guide was used, and a short lever with a small, cast knob was used to shift the assembly into and out of gear. The pulley itself was 13 inches in diameter, made of cast iron. Belt pulley speeds changed slightly at Serial Number 26526, from 948 rpm to 925 rpm.

A view of the top of a Model A. The "muffler" here appears to be a modern straight pipe. Allis had two different mufflers available, both now nearly impossible to find. Allis also offered a straight exhaust stack much like this one.

The underside of the Model A fuel tank, showing the back of the dash. Allis had two different sediment bulbs available, a glass bulb and a metal one. Wiring was probably painted with the rest of the tractor.

The Allis U was born to replace the Fordson. Created by Allis-Chalmers for a group of manufacturers who produced equipment for the Fordson, the U closely followed the Fordson's layout and design, but was a great leap forward in quality. This is tractor number 6076, which had a Continental engine. The air cleaner is either the one used for the last Continental tractors or the later "replacement" Vortox air cleaner listed in the parts book.

CHAPTER 5

Models U and UC

The U and UC tractors share one of the most unusual heritages of any tractors in history. The original design of the tractor was not done by Allis-Chalmers, but instead was done by the United Equipment Company. United was born in 1928, when Henry Ford abandoned U.S. Fordson production. A large number of equipment manufacturers had been supplying the huge market for Fordson implements and attachments, but now were faced with a massive shrinking of their market. Several of these companies formed a group, incorporated it as United, and approached Allis-Chalmers with an invitation into the new company to build a replacement for the Fordson that would sell for less than $1,000. Allis engineers took a preliminary design done by United and completed the design and tooling. The Allis engineers included a former Hart-Parr engineer, C. E. Frudden, and reportedly Charles Parr himself, after he was forced out of Hart-Parr.

In anticipation of the increased business, Allis undertook the construction of a new factory and foundry to lower costs of production and to increase the number of tractors built. The numbers that Allis had built in years past were not nearly what would be possible in the future. Allis produced the U for United (with United on the radiator tank) and also for its own dealers. For various reasons, United fell apart after a short time, leaving Allis with a tractor design that it would produce in one form or another for decades. Many of the attachment manufacturers that initially were members of United continued to build attachments for the U, with manufacturing of the U continuing for the special needs of these manufacturers long after Allis might have ceased construction in favor of new designs.

Variations

Allis offered several variations of the U, for specialized purposes.

UKOs

Allis built "UKO" sets for Number 54 Patrol graders and for Cooper tractor winches. These tractors had U transmissions and other equipment, but had the oil-injected "KO" engine. It is doubtful that Allis sold any complete tractors with this equipment to regular users.

IU Industrials

Available on duals. The Allis parts books list a hand brake being available for these tractors. Bendix foot brakes were an option on IUs.

U Orchards

Several U Orchards were produced at various times. The tractors featured a low drawbar, dash, low seat, and rear wheel fenders. Many other parts had to be altered to accommodate the changes, such as steering gear columns, platforms, gear shift levers, brake levers, clutch pedals, clutch levers, and plates. The air cleaner was also lowered.

By 1936, Allis offered rear wheel fender extensions for use with larger rear tires, including 24x13.5 rears.

UN Narrow Treads

The Us were offered in Narrow Tread versions as well. These used different wheel guards; front axles; dashes; foot rests; clutch rods, levers, and pedals; short rear axles; and narrow rear axle housings.

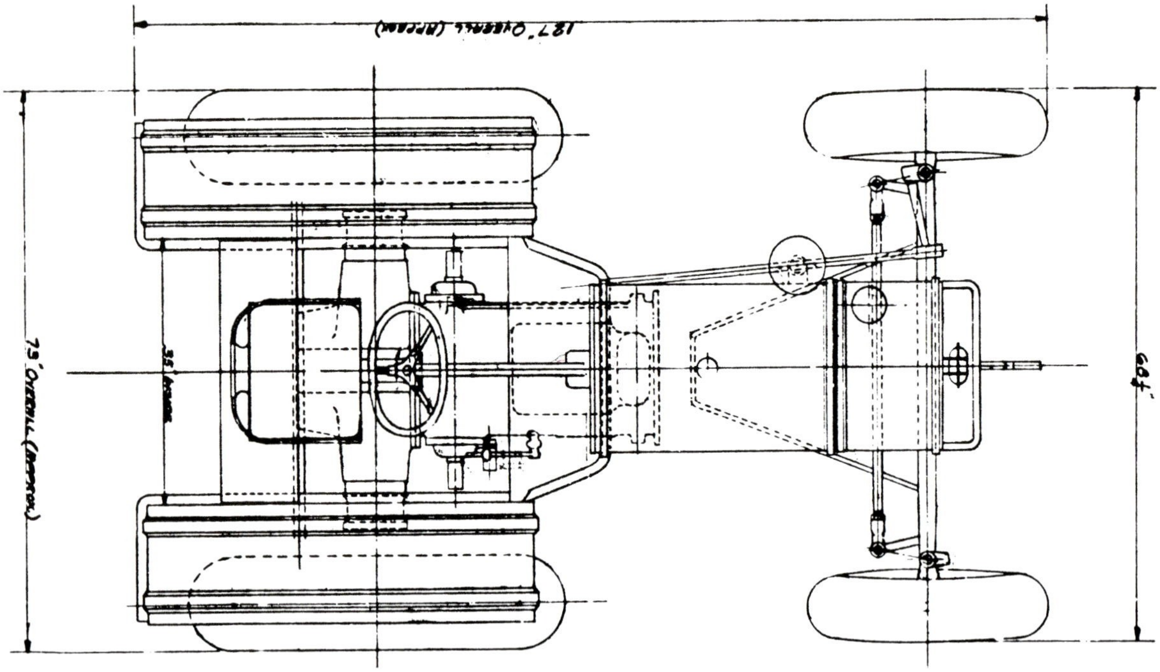

The top view of the UM Cooper tractor winch. Going 30 miles an hour on this tractor, with that short of a wheelbase, must have been exciting.

The U was more or less a transition from an older to a newer style of tractor. The front end was the low-clearance, wide-front "wheatland"-style axle, while only steel wheels were offered at this time for this tractor, built in 1930. The problems with steel wheels were partially solved by the steel overtire seen on this tractor, which allowed movement on paved surfaces. The enclosed platform with sheet metal extending between the hood and fenders reduced dust for the operator in dry western plains farming.

The sign will be familiar to those who have attended the Allis show in Hutchinson, Minnesota, in previous years. While the logo is from the late 1960s (and the sign itself is probably from the 1970s), this 1930 tractor is still going strong next to it. The overtires do give the tractor a forward slant.

UD Diesels

Allis sold at least a few UDs. The tractors had a 2-71 PX-10 Detroit Diesel, and were probably intended for Thompson. The modification apparently lengthened the tractor, as special clutch pedals and other parts were needed. The initial catalog number release for this tractor was in July 1940.

UM Cooper Tractor Winch

Allis-Chalmers made special U tractors for Cooper to produce winch sets for use in oil fields. Two different basic modifications, the 15-mile per hour and the 30-mile per hour, were built. Both of these tractors

U industrials are likely to come in all shapes and sizes. This one was used at Allis-Chalmers' West Allis factory for moving railcars. AC had extensive rail connections and operations, using its own locomotives and railcars to move large pieces of industrial equipment. This U was fitted with a cab to keep crews warm in the winter, and dual wheels to help move heavy loads.

BELOW

This U never received a serial number, and never left West Allis until sold at auction. The large bumper seen here was used for directly pushing railcars. Heavy wheel weights were needed for traction on hard surfaces.

came equipped with pneumatic tires. These tractors were fitted with a 12-inch diameter single-disc clutch. The 1940 tractors used a 7.5x18 six-ply front tires and 11.25x28 six-ply tires on the rear.

Gear Speed Chart for Coopers

15-mph Tractor, 1,800 rpm, 24.5-inch rolling radius.

Allis Catalog Number 290297

First Gear	4.11 mph
Second Gear	5.32 mph
Third Gear	7.42 mph
Fourth Gear	17.49 mph
Reverse	4.69 mph

30-mph Tractor, 1,800 rpm, 24.5-inch rolling radius.

Allis Catalog Number 290298

First Gear	6.75 mph
Second Gear	8.75 mph
Third Gear	12.20 mph
Fourth Gear	28.75 mph
Reverse	7.71 mph

Engine

Three different major engine variations were available for the U family of tractors. The first Us and UCs had Continental engines, easily identified by the upwardly arching manifold. The next engine was an

IUs could be fitted with all sorts of strange stuff from the other companies involved in the United operation. Here's an IU with Allis-Chalmers down the side of the radiator, wheels from Wehr Company, and a hydraulically operated blade from the Frank G. Hough Company (later a part of International Harvester). *State Historical Society of Wisconsin*

Although panned by some other writers in buyer's guides, Industrials have an attractiveness all their own, not to mention a rarity because of hard use. Industrials can often be found with all sorts of interesting equipment. This IU apparently has a PTO-driven hydraulic pump and a reservoir behind it. The air cleaner is the original style of air cleaner used on the Continental-engined tractors. *State Historical Society of Wisconsin*

Allis-Chalmers UM engine with a 4 3/8-inch stroke and a 5-inch bore. This engine was produced for Allis by Waukesha Motors. The last engine was the Allis UM, which had a larger, 4 1/2-inch stroke. This change (occurring either at U tractor Number 12000 or Number 12001) was accompanied by other improvements in the transmission and chassis, probably to handle the increased horsepower.

Engine Serial Codes

Allis provided engines with various compression ratios to ensure that full engine power was available to all users, including those who farmed at elevations where low air densities adversely affected horsepower.

Engine serial numbers are located on the right-hand side of the engine block below the crankcase filler elbow.

K=Distillate under 3,000 feet altitude, compression ratio 4.2:1

G=Compression ratio 4.7:1, gasoline under 3,000 feet altitude, distillate 3,000–6,000 feet altitude

H=Compression ratio 5.2:1, gasoline 3,000–6,000 feet altitude, distillate over 6,000 feet altitude

B=5.5 compression ratio, gasoline over 6,000 feet altitude

Manifolds

Manifolds changed at U-13007 and UC-2700. Before this, low- and high-octane fuel manifolds varied mainly in the gaskets between the bottom of the intake manifold and the top of the exhaust manifold. There were openings in this gasket for low-octane fuel engines, but for high-octane engines, the gasket was solid. On manifolds after U-13007 and UC-2700, a heat control cover was used with a "K" on one end and a "G" on the other.

Carburetors

On the Continental engines, two different carburetors were used— a Kingston L-3 gasoline-only carburetor and a Kingston low-octane fuel carburetor. Zeniths were available as service parts. The Allis engines used the Zenith Z-5 carburetor.

Mufflers

The Continental engines had a muffler that ran horizontally from the manifold to the right side of the radiator. This muffler was special equipment.

The Allis UM engines had two mufflers available. The standard muffler was a rounded saucer-shaped cast-iron muffler with a shield at the bottom. A cylindrical "vacuum type" muffler was also available.

Exhaust Horn

A Fulton exhaust horn was available for the Model U. It attached to the muffler, and was a four-chime unit. A cable ran from the unit to a pull knob near the operator. Allis' part number was U-1026.

Magnetos

The Continental engines used a base-mount magneto with a tube assembly for the spark plug cables. The base-mount magnetos used included Bosch, Eisemann GL-4, and Scintilla. In later years, Allis offered only Fairbanks Morse magnetos for service parts. The cable tube assembly was provided by Continental Engines under the part number S10M-500, and appeared to be a piece of thin-walled tubing bent to shape.

This Model U tractor has an IH magneto, which isn't original. Many of the magnetos that Allis used went out of production in the 1930s or 1940s, and more common (and sometimes more reliable) replacements were fitted.

Different coupling drives for each magneto were necessary. The coupling drive was completely exposed.

The first UM engines (before engine UMA-4587) also used a base-mounted magneto. The base was a curved bracket. Eisemann GL-4, Bendix C-4 (discontinued on engine UM-4587), and Fairbanks Morse RV-4 and FM04A magnetos were used with several different couplings. The bracket hid most of the governor shaft.

On engine UMA-4587 and after, a flange- mounted magneto was used. Fairbanks Morse FM-4, RV-4, and FMOR magnetos, Eisemann GL-4, and Scintillia C-4 were all used.

Oil Filters

The first oil filters used were Purolators with an internal element. The UCs with Continental engine had a slightly different bracket than the Us. The can-type oil filters were used from 1937 onward. The oil gauge was built into the base unit of these filter assemblies.

U Continentals used a Purolator oil filter with an internal replaceable element. Purolator supplied the instruction decal that went on the exterior of the canister. Many makes of tractor used similar oil filters, and the decals were the same across the makes. The tap at the bottom of the filter is to drain the canister itself during a filter change.

Governor Controls

Before U-12001 and UC-2819, a governor control rod that had a rack machined into it was used. The rack rested on the dash, with the edge of the dash engaging the rack, preventing it from moving. To change engine rpm, the operator just lifted the rod, moved it in or out, and then put it down on the edge at the bottom of the slit in the dash. The IU used the same system. The rods were different lengths, depending on whether the tractor was a U or UC, and whether the engine was the Continental or UM.

After U-12001 and UC-2819, a quadrant and ratchet system was used. This same system was used in the preproduction Bs and As.

Air Cleaners

Tractors with U and UC Continental engines used a Donaldson air cleaner with centrifugal airflow and a dry element. Tractors before U-5872 used a glass dust Mason jar type, while later tractors used a more traditional precleaner mounted on the intake stack. These tractors also had a different cleaner casing than the later Continental Us. Orchard Us had a different mounting bracket. Allis offered a replacement oil bath type air cleaner for the Continental Us.

Vortox air cleaners were used in early production Allis UM-engined tractors. Donaldson air cleaners became standard at U-13007 and UC-2700. The Donaldson air cleaners on tractors before U-16848 used a combination bracket and air intake, while Donaldson air cleaners on tractor U-16848 and after used a separate mounting strap and a rubber outlet hose. These tractors also had a glass jar type of precleaner.

Transmissions and Accessories

A transmission boot, 205333, was used on tractor U-12000 and after. This coincided with the use of a new transmission cover. Five different shift levers

Chassis Model U Dimensions, from the Advertising Literature

General Dimensions—Wheelbase, 76 1/2 inches; overall length, 121 3/4 inches; overall height, 65 inches; overall width, 66 inches; center-to-center tread, steel wheels, 55 3/8 inches; center-to-center tread, air tires, 55 5/8 inches; ground clearance, 11 inches.

Standard Equipment—Steel wheels or air tires as ordered, clutch-type belt pulley, rear wheel fenders, large enclosed platforms, large seat with upholstered seat cushion, thermostat, radiator curtain.

Extras—Oversize front and rear tires, additional rear wheel weights, front wheel skid bands of various heights, angle-iron cleats, 6-inch spade lugs, PTO, electric starting equipment.

were used, as were two different shifter knobs if you see something that your neighbor's tractor doesn't have, don't be surprised.

Clutch controls varied between tractors and years. The original system was a step-down pedal—the operator stepped down on the pedal to disengage the clutch. The Us before Serial Number U-9245 used pedal U-166. U-1271 was used between tractors U-9245 and U-12001. U-745A was used on U Orchard tractors with Continental engines. U-797 was used on narrow-gauge tractors. U-1105 was used on all UCs until the end of production. Narrow-gauge and orchard tractors had their own clutch rods. The clutch rod assembly adjusting mechanism was changed at U-9245.

On Us with Serial Number U-12001 and after, a push-forward clutch pedal was used. This utilized a curved pedal arm with a pedal assembly that could be adjusted in and out to suit the operator. IUs also had a push-forward type of clutch pedal, but instead of the curved multiple-piece system of the U, a long, straight pedal assembly was used.

On Us and UCs with steel wheel equipment, fourth gear was locked out.

Radiators

"United" U tractors often had the word "United" cast on the top radiator tank. However, as the accompanying photographs prove, some tractors were sold by United that had "Allis-Chalmers" on both the sides and the top of the tank. The diamond on the side of the radiators never had the "U" cast in, as the Allis crawlers did.

The top tanks changed in one other way. Early top tanks had a round filler hole, with a round cover. Later tractors had an oval hole and a hinged cap. Radiator curtains were available, and were made of olive drab cotton duck. A radiator grill guard was also available.

Canopies for Us

Part U-1072 was a canopy for the U tractors. The canopy was made from steel bar and angles, with a brown canvas duck covering.

Rear Fenders/Platforms

The U had an enclosed platform typical of the standard tractors produced in the 1920s and 1930s. The footrest used on the first 1,000 tractors was U-235, a triangle-shaped plate with the top corner cut off. Footrest U-372 was used before tractor U-7405, and was a straight flat bar 9 5/16 inches wide by 1 1/4 inches high. The next footrest was part U-1989, which was also a piece of steel flat bar, 24 inches wide, with both ends bent. U-1989 was used from tractor 7405 to 11999. Finally, for the 1936 season tractors, the footrest was changed to "floor plate" 204118 on

This view shows the torque tube construction of the U, although when this tractor was built, it was called a "clutch housing." The air intake pipes and oil filter cap were probably painted orange by Allis. The clutch lubrication decal is seen just above the draglink in front of the overtire.

This is what is known as the "winged" gas cap, found on some 25-40s, U, UC, and WC tractors. The wings were easily broken off, and finding original unbroken caps is difficult. However, the caps are now being reproduced.

In this view of Larry and Edwin Karg's Model U, the Allis diamond on the rear of the fuel tank, the seat construction, and the platform can be seen. The gear shift knob was, in all likelihood, painted orange by Allis. Although there was no paint specification for the steering wheel, it also could have been painted orange.

tractor 12000 and after. This "floor plate" was a complicated piece of sheet metal.

The platform was extensively changed at U tractor Serial Number 12000. The toolbox was mounted on the left fender on all the U tractors.

The two UC rear fenders were the same part number, U-689, before tractor Serial Number UC-2819. Construction was changed on UC-2819 to a right-hand and a left-hand rear fender. The toolbox went on the left fender.

Fuel Tank/Dash

Although the main fuel tank itself seems not to have changed over the years, everything else connected to the area did. Fuel tank caps, for example, changed from an older brass design with the built-in "wings" to a round cap with ridges pressed into the cap. The front support of the tank on the U Continentals was different from the front support for the Allis-engined tractors.

The rear dash/fuel tank support changed several times, and varied by model as well. On the standard Us, the dash was changed at serial U-12001. The UCs before UC-2819 also had a dash different from the later tractors. IU Continentals had their own separate dash, which changed with the introduction of the Allis engine. U Continental orchard and narrow-gauge tractors had their own dashes, but seem to have used the standard U dash after the introduction of the Allis engine. Allis-engined UCs and IUs with electric lighting also had different dashes from their lesser-equipped siblings.

A sheet-metal clip fastened to the engine block between the sediment bulb and carburetor to hold the fuel lines. Both metal and glass bowls for the sediment bulb/fuel filter assembly were available. The sediment bulbs for the low-octane fuel tractors had two inlets for fuel, while the gasoline versions had only one inlet.

The auxiliary gasoline tank changed with the introduction of the Allis engine. There were two assemblies, U-2027 and 226233. The diameter of the fuel line and shut-off tube was decreased from 5/16 to 1/4 inch.

Hoods

Hoods were shared between the U Continental and the UC Continental. One hood was applied to gasoline tractors, while another hood had the necessary extra cutouts for the small gasoline tanks for the low-octane fuel tractors.

The UCs before serial U-12001, Us before serial U-2700, and IUs before serial IU 10765 had common hoods, whether low-octane fuel or gasoline. After this, U tractors U-12001 to U-13007 had a unique hood (205804). U tractors 13007 to U-16848, UC tractors UC-2700 and after, and IU tractors U-10765 and after all shared hood 211905. At least some IUs received hood side plates. These tractors were fitted with yet another hood assembly, U-1824, which was fitted with the attaching hardware for the side plates.

Electrical System

Electrical starting was available from the beginning on the U and UC. Motors and starters were Auto-Lite. On the Continental engine tractors, the ammeter was held in an instrument bracket that constituted a small panel. A push switch was mounted on a starting switch bracket attached to the steering box mount. Two different switch and bracket combinations were used. The first UM engines used a different instrument bracket, with the starting switch being mounted in the dash of the tractor. On the second type of arrangement on the UM engine tractors, the ammeter was located in an instrument box.

Lighting systems on early Us with batteries were either 6-volt or 12-volt, although over time, the 12-volt system seems to disappear. On the early "first type" lighting systems, large, flat, cylinder-type headlight housings were used. Two headlights and a taillight were used, and the switch was mounted in a sheet-metal bracket. The bracket on the IU Continental was different from the U, UC, or IU with the UM engine.

The "second type" lighting system with battery used teardrop-shaped bulb housings (made by Guide Lamp) for the headlights, and a small cylinder-shaped housing for the taillight. The light switch was held in the instrument box that also held the ammeter.

Two different batteryless electrical systems were used. The first type used a Bosch generator system, with a turn switch located on the generator itself. The second system was from Auto-Lite, and featured a pull switch located on the generator.

The UC was the row crop equivalent of the U. The tractors were designed from the start to use common components for the very different jobs of standard tractor and row crop tractor. This photograph is of one of the early "All Crop" tractors with a Continental engine. *Photo courtesy of Norm Swinford*

Model UC Dimensions from the Advertising Literature

Overall Dimensions—Wheelbase, 87 1/4 inches; overall length, 125 1/8 inches; overall height, 67 5/8 inches; ground clearance, 30 inches under rear axle housing.

Standard Equipment—Steel wheels or air tires as ordered, clutch-type belt pulley, hand brakes, rear fenders.

Extras—Tire chains; additional wheel weights; angle iron skid bands for front wheels in various heights; 6-inch extension rims for rear wheels inside and outside; 4 3/4 x 5 spade lugs for extension rims; 3 1/2 x 3 1/2 x 13 5/8 angle iron cleats, used only with two extension rims per wheel; PTO; power lift; cushion seat; electric lighting equipment; furrow guide.

Front Axles and Steering

The standard front axle for the U and IU was a solid bar type. The IU had an optional front axle with leaf springs, and a wider 44-inch front axle for heavy-duty applications. The UC had a standard two-wheel narrow front axle. The very early UCs (before UC-32) had a different front axle support and assembly. U-400 was a special high-strength front axle made for the United tractor out of nickel steel. The letters "NS" were pressed into the axle to identify it.

The U and the IU tractors used a steering system that included a gearbox and a draglink assembly. A Hanum steering gearbox was used on Us before U-12401 and IUs before tractor U-10002. U-12401 through U-13501 and IU tractors 10002 through 10425 used a Lavine steering gear. On Us from U-13501 and after, and IU 10425 and after, Ross steering gears were used. The Hanum and Lavine gears had a parts availability problem, and the Allis parts manual stated that if parts were needed other than the steering wheel, steering gear covers, and bolts and dowels, that a complete Ross steering gear assembly had to be purchased. The Ross assembly that replaced the previous units was different from the original Ross unit. The agricultural As used a longer cam and shaft assembly in the steering than the industrial As (36 1/32 inches versus 26 25/32 inches). The tube that the worm gear rode in was also different.

UCs had a simpler steering system, involving shafts running from the steering wheel to the front axle support, which enclosed the gearbox. All Us used steering wheel U-141.

Brakes

As noted in the variants section, both hand and Bendix foot brakes were available on IUs. For the Us, a pedal brake lock was used on tractors before U-1008, and a hand lever lock used from U-1008 to U-12001. The brake levers themselves changed at U-12001, but U Orchards used their own brake lever (U-675) throughout production.

PTOs

The PTOs for the U and UC tractors were very similar. The UC used a different intermediate shaft. The actuating levers changed in 1936. A long lever was used before 1936. In 1936 and after, a short lever shaft was used. After 1938, a shifter rod was added to the short lever.

A PTO guard, U-1243, and a flange guard, U-1244, were used before 1940. After this date, guard assembly 210481 was used.

A double PTO shaft unit was available for the IU tractor. The unit had a gearbox that attached to the rear of the tractor and drew power from the regular PTO shaft. The twin shafts were located one above the other, and were several inches above the regular PTO shaft location.

A sheet-metal PTO guard with a cast bracket was used until approximately the start of 1940, and after that time an all-sheet-metal guard with braces (and warning decal) went over the top of the end of the shaft.

Belt Pulleys and Rollers

Due to the layout of the tractor, a belt pulley roller was used on the Us to guide the belt over the front axle. The roller attached to the right front side of the tractor.

Belt pulley assemblies changed several different times, and had several different configurations. The three major configurations were belt pulleys without clutches, straight belt pulleys with clutches, and elevated belt pulleys. The belt pulley without a clutch underwent a major change at Serial Number 5951. Tractors before this had a straight shaft on which the belt pulley was mounted, while 5951 and after had a tapered shaft. The tractors with straight shafts had an iron pulley, while tractors with the later tapered shaft had an iron pulley or a Rockwood pulley.

The regular belt pulley with clutch had the same pulleys available as the tapered shaft pulley without clutch. Minor mechanical changes were made to the assembly over the years. The clutch pulley was actuated by handle U-2911.

On tractors U-12001 and after, the elevated belt pulley with clutch was used. This assembly was similar, but not identical, to the belt pulley assembly on the A tractors. The pulleys were the same ones used on the other tapered shaft assemblies.

Power Lifts

Two different power lifts were available for UC tractors to assist with lifting cultivator gangs—a light-duty unit and a heavy-duty unit. Both were worm gear driven and pedal operated, although the light-duty unit could be fitted with a lever for hand operation.

This is a slightly later UC, fitted with skeleton-type rear steel. Allis provided tractors with a wide variety of tire and wheel equipment. The UC was probably the first row crop tractor to be offered with rubber tires. *Photo courtesy of Norm Swinford*

This shows the Model U belt roller in reference to the front axle and wheel. Us used a variety of wheels and tires over the time of production, with some tractors leaving the factory without wheels or tires. Customers in foreign locations, or domestic locations with unusual conditions or soils, would fit local equipment.

Early Us used a belt roller to clear a flat belt over the front axle. The original design of the U had a normal belt pulley drive located low on the tractor. Later Us had an offset belt drive unit that moved the belt pulley higher than the centerline of the drive.

LEFT
As mentioned previously, the diamond on the side of the radiator does not get a decal, nor was there ever a letter cast there the same way that there was for the Allis crawler line. The UC was one of the first general purpose tractors to receive air tires. In 1938, the color of the rims and wheels would have been black.

This very nicely restored U dates from 1938. Allis-Chalmers continued to produce the U as its large row crop tractor, even after the much more advanced and popular WC and B entered production.

Visible on either side of the UC steering wheel are the brake levers. The brakes were used to help steer the tractor, especially at the end of rows. Allis changed the paint specifications for steering wheels several times. Many wheels were painted all orange, including the rim, but some rims were left black by the steering wheel manufacturer.

The position of the AC diamond decal is clearly shown here. The U and UC (along with the A) were never "styled" or streamlined to the degree that the other Allis tractors were, probably due to their low production numbers. The result was that these square, flat surfaces were produced far longer than unstreamlined tractors from other manufacturers.

The right side of the engine mount on the UCs shows the spark plug wire cover and the magneto. The magneto is a Fairbanks Morse, and is flange mounted.

ABOVE LEFT
The UC had large fenders compared to most of the row crop tractors of the day. The rear tires have a 45-degree angle tread pattern, which was just coming into use in 1938. The new lug pattern further improved tire performance over the original pneumatic tires, which themselves were an improvement in performance.

LEFT
The UC does not seem to have been available from Allis-Chalmers with a wide front end, although wide fronts were available from companies who built tractors from UC components, like Thompson.

The Us before Serial Number U-12000 used a seat support mounted on a bracket, with a metal seat pan and a seat cushion. Orchard models used the same parts with a seat brace and a spring. After Serial Number U-12001, a seat with a back was used with a completely different mounting.

The UC still had the old fashioned bail-type radiator cap, held down by a wire. On the right side of the hood, near the front, is the gas cap and the attachments for the auxiliary gasoline starting tank. The air cleaner on UCs of this vintage did not have a cap or a precleaner, resulting in the second open-top pipe at the left front of the tractor.

RIGHT
By 1938, the UC had the Allis-Chalmers UM engine, instead of the earlier Continental engines. The manifolds were a completely different shape. The air cleaner decal is missing on this example.

This UC tractor has the U-3550 seat that was used after serial number UC-2819 (this tractor's serial number is UC-3874). The seat has a back and a cushion. The Allis diamond decal is in the correct place, in the middle of the back of the fuel tank. The drawbar construction also changed to include a bend near the end at UC-2819, as shown on this tractor.

The WD was put into production in 1948 as a major upgrade of the WC. Improvements included more horsepower, foot brakes, and a "transmission" clutch that allowed the transmission to be thrown out of gear while the engine clutch was left engaged, supplying power to the PTO.

CHAPTER 6

Models WC, WD, WD-45, and WF

The next tractor family designed by the Allis engineers was the WC, targeting a lighter row crop market than the UC. The basic design lasted through a large number of sheet-metal, drivetrain, engine, clutch, brake, and other changes, becoming first the WD and then the WD-45. Throughout its variations, it was produced from 1933 to 1957. The Allis WC was very light for its size and horsepower, and proved to be an extremely popular tractor for many years. In addition to the row crop versions, Allis also produced a standard tractor with the same engine and driveline, the WF.

Some WCs were produced for cane conversions for Thompson, while others were produced for sale to regular customers. The Thompson tractors had an arched front axle that yielded 26 1/4 inches of ground clearance. WC Cane units were fitted with 4-inch-wide by 26-inch-diameter steel front wheels (part number 213499). The wheels and axle gave a front tread width of 66 inches. The rear wheels were 6 inches wide by 54 inches in diameter, and the gear ratio of the rear axle (parts group 281805) was 6 3/5 to 1. WC Cane models for regular customers used 9.00x36 or 10.00x36 rear tires.

The WD was a vastly improved WC, the primary changes being a new transmission clutch that allowed forward travel to stop, while maintaining power to the PTO. This was used to great advantage with the All-Crop combines and Allis Roto-Balers. The feature had been born in the D and F tractors that Allis had attempted to put into production starting in 1940. Unfortunately, World War II interfered at first, followed by strikes and other postwar problems, so the upgraded WD was produced instead. The WD also featured foot brakes. The WD was available with three front-end configurations: two-wheel narrow front, wide adjustable axle, and a single-wheel front. The WD also featured the new Power Shift rear wheels, which allowed an operator to change the rear wheel tread width by merely loosening a few bolts and driving forward or backward, a great advance over previous methods where the entire wheel and tire (full of fluid) had to be moved in or out manually. WDs were produced in cane specification and for Thompson. They also included a cotton picker version. An LPG fuel version probably became available in June 1952.

The WD-45s were introduced in 1953, with starting Serial Number WD146607. The first WD-45 Diesel was Serial Number 181341. The WD-45 was available in three different front axle configurations: the narrow front with two wheels, the single-wheel front, and the adjustable wide front axle. Gasoline, LPG, and diesel versions were available. Options for the WD-45 included detachable seat pad 226173, rear wheel weight 222943, front wheel weight 223572, and belt pulley 282389. WD-45s were also manufactured in cane versions.

Engines

The first WC tractors, built in 1933, used Waukesha engines, easily identified by their flat heads. After 1934 the WC used an Allis-designed engine (with some work still being done by Waukesha).

Gasoline-only engines were fitted with a "cold" manifold without a cover. Kerosene ("two-fuel") engines were shipped with two manifold covers. One manifold cover was marked with a K for kerosene burning, while a manifold cover for gasoline burning was marked with a G.

continued on page 72

This WC was built in 1944, at the height of World War II. There are many minor material changes in several parts, and the original tires were probably made of synthetic rubber, which had a certain whitish appearance.

Some parts of the WC were used on both the styled and unstyled tractors. Seats were the same, as was the engine, driveline, and steering. Unstyled WCs used square-top fenders, while styled WCs used curved "clamshell" fenders that flowed over the rear wheels.

The "styled" WC picked up where the unstyled left off, in terms of sales, popularity, and usefulness. The WC was the tractor that made Allis a leader in the farm tractor industry. The smooth lines of the tractor, first appearing in 1938, put Allis among the first companies (after Oliver and Graham Bradley) in styling their main row crop tractors.

The major problem with the WC was the hand brake. Row crop tractors tend to make fast turns at the end of rows, using the brakes to prevent the tractor from skidding. Use of a hand brake, while at the same time trying to move the steering wheel, led to some exciting adventures with fence lines. The stowage for the starting hand crank can be seen against the fender.

When many people see something like this, they think they should disassemble the tractor down to parts and reassemble it into something "original," i.e., the way it came off the Allis assembly line. However, a conversion like this, far beyond what Allis did with their high crops, is inherently rare and more than worthy of preservation.

The unstyled WC tractor used for the conversion was made in 1935. A lot of restorers leave the radiator top tank unpainted, but it was painted by Allis. A more modern oil filter has been used on this tractor. The correct seat is there. Wouldn't you like to crank this tractor? Usually you have to keep your thumb out of the way—on this one, you've got to keep your head out of the way.

Unfortunately, neither the owner nor the author knows much about this particular conversion. It's pretty spectacular, but clearly factory made. The decal on the radiator doesn't seem to be anything Allis used in production. (Early WCs did use a decal on the radiator, but may have been confined to the very early Waukesha-powered WCs.) A later oil filter, of course, is used.

RIGHT
The rear drive is completely non-Allis. The exposed chain drive must have been a maintenance problem—other manufacturers used chains for the same purposes, but put them inside a housing. The tires are vintage 45-degree treads, as opposed to the more modern 33-degree treads that put the lugs at shallower angles.

The WD packed quite a bit of power for its size, as well as a nice range of features. The tractors can do quite a bit of work, including pulling this drag on oats planting day at the Rock River Thresheree. This WD has an orange steering wheel hub with a black rim. Allis steering wheel colors changed several times.

The trade name decals on the frame rails are close to where the originals were placed. The seat is the proper type for all unstyled WCs, although there were several different types of seat mountings. Somehow, I don't think this tractor was intended for dairy farms, unless it was used to pick up cows and carry them underneath.

RIGHT
The WD also featured the new "Power Shift" rear wheels, seen here, that had inclined rails built into the rim. The bolts could be loosened, and lugs on the wheels acted on the rails, forcing the wheels in and out with the power of the tractor, to change tread width.

This WD is missing the hydraulic pump instruction decal that was placed on the right side of the battery box. The oil filter is a modern replacement. The radiator shutter instruction decal is missing and the tires are the more modern version with the 33-degree-angled lugs. Orange front wheels seem to be pretty much the standard for tractors of this age (1951), although some front rims may have been silver.

ABOVE
This archival photograph shows a WD-45 Diesel. This tractor still has a black generator, although the rest of the left side of the engine seems to be orange. The tire instruction decal can be seen on the left wheel at the upper left of the hub. The taillight on this tractor shows up as orange. The steering wheel rim is black, but the wheel hub and spokes are orange. *Photo courtesy of Norm Swinford*

LEFT
The right side of the WD-45 Diesel shows the hydraulic decal on the right side of the battery box, and the different manifold system used on the WD-45 Diesels. The "Diesel" is below the "WD-45" decal and is actually a separate decal. Note the two sets of wheel weights on the rear wheels.

Although the WD-45 is similar to the WD, the extra horsepower coming from a slightly larger and redesigned combustion chamber produced a whole new range of performance. Going from a WD to a WD-45 in plowing was like going from a family sedan to a musclecar. Look at all the weights on the tractor in front of the radiator!

The "WD" decal on this tractor doesn't quite look original. The decal placement on the engine block cooling drain (seen to the right of the generator) is about right, as seems the placement on the "United" decal on the air cleaner. The carb and air hose would be painted orange on an original production tractor. The color of the generator and starter probably would have matched, although whether it should be black or orange might be a toss-up as the colors did seem to change at about this time.

On original production tractors, the coil and distributor caps would be painted orange. However, since these items were replaced quite often, and the replacements were usually black, in very little time an almost-new tractor would have them black. Allis actually supplied oil filters in original colors and decals for quite some time, but the logos changed long ago.

This is an archival photograph of a late WC. This WC has orange front wheels and rims, and orange wheels with silver rims on the rear of the tractor. The magneto on this tractor is painted all orange. Note the position of the trade name and radiator shutter crank instruction decals. *Photo courtesy of Norm Swinford*

The left side of the WD-45 engine shows that the generator has been replaced by the alternator. Collectors commonly find this replacement. Alternators, essentially car parts, generally have more charging capacity, and in later years were much cheaper than replacement generators. Tractors with 12-volt electrical system conversions will often have alternators—they're usually shorter and wider than generators. This collector has painted his air cleaner black, which is not original, but it sure draws attention.

The WD-45 block did not have the large access plate on the right side that the WD had. This tractor has an original-looking oil filter, the "Power Crater" engine decal (describing the new shape of the combustion chamber), and black headlights. The distributor cap and coil would have been orange on a freshly produced tractor.

The extra length of the WD-45 Diesel can be seen in the "stretched out"curve of the front hood. The exhaust and air intake cap are not original, but the rest of the tractor looks pretty good. This tractor was built in 1955, the same year that many of the Allis parts prints were changed to specify that the steering wheels be all orange. The rear blades are a very handy thing to have—the author's father had a three-point Allis blade that saw use summer and winter.

Continued from page 65

WD-45 gas engines used a block that did not have the large pushrod cover plate on the right side of the engine, as the WC and WD engines did. WD-45 engines were considerably more powerful than the previous WD engines, using a hemispherical "Power Crater" piston and a higher compression ratio.

Engines with Serial Number 299173 PA and after were fitted with exhaust valve rotators, although the rotator caps could be fitted on all W series exhaust valves. Because the WF tractor stayed in production several years longer than the WC, it received engine upgrades with WD engine parts, including the rotator cups and cylinder heads.

Engines before Serial Number 274626 had what was called the "first type" sump with the oil drain on the side of the oil pan. After engine 274626, the oil drain was on the bottom of the pan. On WD and WD-45 gas and LP tractors, the oil gauge was located on the oil filter boss near the engine. On WD-45 Diesels, the oil gauge was part of the three gauge instrument cluster on the top of the hood.

The WD-45 Diesel's longer engine resulted in less of a "platform" between the front of the pedestal and the front of the grill. The complicated injection pump piping can be seen here. The complexity, high fuel pressures, and tight tolerances of the diesel fuel systems often prompt restorers to send these parts out to qualified injector shops for service.

The WD-45 Diesel was the first major Allis entry into the field of diesel ag wheel tractors. The tractor used a six-cylinder Buda diesel that greatly lengthened the tractor. Slightly over 6,500 were built, well under 10 percent of WD-45 production.

On the right side of the WC, the "grease universal joint" decal goes on the steering shaft about 1 1/2 inches below the support seen here in front of the headlight. The radiator shutter crank instruction location is pretty close to original. The oil filter is a modern replacement, and the magnetos tended to be painted in production.

On this WD-45 Diesel, the red lubricating oil filters seen next to the frame rails are modern replacements. Diesel tractors also have elaborate fuel filtering systems. The injection pumps and injectors have very fine tits and tolerances that are easily damaged by dirt in the fuel system. The primary fuel filter system is under the fuel tank. There is an intermediate and secondary fuel filter located near the cylinder head under the hood at the top of the curve.

WD-45 Diesels used a Buda (which had by then been purchased by Allis) six-cylinder direct start diesel engine. The extra length of the engine required different control linkages, frames, and hoods. The company's original experimentation on diesels had taken place in WD tractors, with engines purchased from Buda. All parts of the Buda diesel are different from the Allis production gas engines.

Oil Filters

The original W engine oil filter was a Ball Brothers pint glass jar. The jar screwed into the filter base and was filled with filter material. When the oil in the jar darkened, it was time to change the oil and the filter material.

In November 1935, the jar was replaced by a can-type filter (part number 204862) that had the Allis logo, part numbers, and patent numbers lithographed on it, with the filter material integral to the can. On the top of the can, the words "Replace filter when crankcase oil darkens. Ordinarily life of filter 100-200 Hours" were centrally lithographed, and later replaced by a decal. A patent number was added in August 1941, and the entire can design was changed in September 1943. Patent number 2287982 was added in September 1947, and in September 1948, the lithography on the can went through another major revision. In April 1949, the part number on the can was changed from 204912 to 800912. The can used black lettering on orange enamel.

Engine Serial Number Markings

Allis provided engines with various compression ratios to ensure that full engine power was available to

WD-45 Diesel Dimensions, from the Advertising Literature

Dimensions—Wheel base, 88 1/8 inches; overall length, 128 inches; height at muffler, 88 1/4 inches.

Standard Equipment—Starter and lights, hydraulic system with Traction Booster, Snap Coupler hitch, Power Shift wheels, Two-Clutch Power Control, ASAE drawbar, PTO.

Extra Equipment—Belt pulley, remote hydraulic ram, front wheel weights, rear wheel weights.

WFs were used for more than farming. This tractor has a Drott dealer plate on it. Drott manufactured industrial equipment and ran some large industrial equipment dealerships in Wisconsin. The plate suggests that this tractor was originally sold for industrial purposes.

all users, including those who farmed at elevations where low air densities adversely affected horsepower.

K=4.2:1 compression ratio
G=5:1 compression ratio
B=5.5:1 compression ratio
G=5.5:1 (W101809 and after)
GA=5.5:1 compression ratio

WD Engines

KA= Kerosene
PA=High compression gasoline

Manifolds and Mufflers

On WC and WF tractors, for all low-octane fuel and gasoline engines before engine W-106509, a manifold with the runners running underneath the centerline of the manifold was used. The central "box" cast

The first 1,904 WFs were produced in an unstyled version, similar to the unstyled WC. After these first few tractors, Allis switched to using the styled parts for these tractors. WFs were never produced in numbers like the WC, although WF production extended three years beyond the end of WC production in 1948. This WF was produced in 1950.

The WF front axle and spindles had nowhere near the height that the WC axles had, and the rear axles didn't have the final gear drop housings. However, many other parts, including most of the drivetrain, were produced with the same parts used in the WC.

The WF was the "standard" version of the WC, created to market a tractor to farmers who didn't need a row crop tractor, but didn't need a tractor as large or expensive as the A. Crop clearance and height were much reduced.

into the manifold could be fitted with two covers—one for gas, one for low-octane fuels. The exhaust was angled out from the manifold, not going straight up through a hole in the hood. The standard exhaust looked like a piece of pipe flattened on one end. A muffler was available as an option, which was short and had a wide, flat oval chamber.

WC and WF gasoline tractors with engine serial numbers after W-106509 have a manifold with runners both above and below the centerline of the manifold. These tractors have a muffler that goes through the hood. The muffler is a large diameter round can-type that sits very near the hood, with a very short pipe section on top.

Low-octane WD and WD-45 tractors had different manifolds, but had the same box-type manifold covers before WD-45 Serial Number 161825. Low-octane tractors had a manifold/manifold cover assembly similar to the WC. Tractors before this serial number had an angled exhaust coming out of the cover, and either the crimped pipe-style exhaust muffler, or the short angled oval can-type muffler. After WD-45 Serial Number 161825, the manifold cover had the exhaust pipe sticking out the top through a

The WD-45 Diesel's Buda diesel used a manifold that had an enclosed air heater for preheating intake air. This was necessary to start the engine in cold weather, since diesels rely on heat for fuel ignition. The preheat chamber is the square box at the junction of the intake air hose and the intake manifold.

This WF tractor has a black generator with masked-off data plate and an orange carburetor. The "Allis-Chalmers" decal seems well aft of the original position, and the air cleaner is missing its decal, as is the battery box top.

The WD featured a double quadrant, mounted on the steering wheel support, that had control levers for the hydraulics (left side) and the throttle (right side). The quadrant was painted in production, and the tips of the levers had rubber knobs. A hydraulic lever instruction decal would have been placed on the horizontal straight crossbar on the left side of the quadrant.

One of the major differences between the unstyled and styled WC is the enclosed radiator. A sheet-metal housing and grill both protect the radiator and change the shape into a "streamlined" tractor, as opposed to the squarer unstyled tractor. The two hand brakes can be seen here, next to the fenders.

hole in the hood, and a round narrow can muffler with a bottom flap to attach the muffler to the manifold cover.

Gasoline WDs had a one-piece manifold with a plate-type manifold cover. The exhaust was vertical, using a modern-type muffler. One manifold was used for gasoline-fueled tractors, while an LPG/Butane manifold was used for those tractors. A muffler clamp and muffler support were used on tractors Serial Number 17786 and up. Tractors before this used a bolt.

WD-45s used a one-piece manifold without a cover and a modern-type muffler with a bottom flap to attach to the manifold. WD-45 Diesels used a flap-type muffler, with a cast exhaust pipe running between the manifold and the muffler. The diesel engines also had an air heater chamber located under the manifold.

Carburetors

Unstreamlined tractors (before WC-74330 and WF-1904) used Zenith 124 carburetors. Streamlined WC and WF tractors used Zenith 61 or Marvel Schebler TSX-159 carburetors. The parts books refer to one carburetor assembly, Allis part 219199, as being for alcohol use.

WD low-octane tractors used either Zenith carburetor 213348 or Marvel Schebler 222089. WD-45 low-octane tractors used a Marvel Schebler carburetor. Marvel Schebler and Zenith carburetors were used on regular gasoline versions.

WD and WD-45 LPG tractors used an Ensign C-2588 regulator; WD tractors used an Ensign 2554 carburetor, while the WD-45 used an Ensign 3060 carburetor.

This WC has a "United" type of air cleaner decal. The "WC" decal is from one of the unlicensed decal manufacturers—no such decal was ever made by Allis during the production of the WC. The generator is black and the carburetor orange, as were the originals. It is not known why the air hose isn't connected.

Magnetos

WCs with serial numbers before W-3522 used a base-mounted Bendix Scintilla C-4 magneto with exposed coupling. One cable tie was used on the spark plug wires. The ground is located on the magneto itself.

WC tractors W-3522 and up, and all WF tractors, used a flange-mounted Fairbanks Morse magneto with enclosed coupling. After the Bendix magneto became unavailable, Allis offered kits to convert tractors to the Fairbanks Morse unit. One cable tie was used on the spark plug wires. A grounding switch (to shut off the engine) was located near the operator.

Fairbanks Morse magnetos were standard or optional in WD and WD-45 tractors after number 136318. The magneto for the LPG or Butane tractors was slightly different from the gasoline models. A cable tie was used on the plug wires.

On Serial Number 136318 and up, a distributor ignition became standard. No cable tie was used on the plug wires with the distributor.

Governor Controls

All WCs and WFs before tractor 1904 used a quadrant with "high" and "low" cast into the part. On WFs number 1904 and after, a squarer quadrant was used without the wording.

WD and WD-45 tractors used a combination hydraulic/throttle quadrant located on the steering shaft. The WD-45 throttle was fitted with a friction block. The end of the knob was fitted with a rubber knob. WD-45 Diesel tractors had a fuel shut-off latch built into the throttle lever.

Cooling

Prestreamlined WC and WF tractors used a radiator assembly with Allis-Chalmers cast into the top tank. The radiator had a screen assembly and a radiator curtain (made of olive drab cotton duck). WC-29845 and after used a water gauge that was inserted in the water inlet on the top tank.

Streamlined WC and WF tractors had both pressurized and nonpressurized radiators available. The water temperature gauges and radiator caps came in pressurized and nonpressurized versions to match. The olive drab cotton duck radiator curtains were used to Serial Number WC-101353 and WF-2303, when they were replaced by a shutter assembly. The radiator was covered by a shell assembly that enclosed the radiator, and included the grill screen, to streamline the tractor. The shutter crank extended through the right side of the shell, which had a shutter instruction decal located below the crank.

WD and WD-45 gasoline and LPG tractors shared radiators, but WD-45 Diesels had a separate radiator. The radiator core assembly had two pieces of angle attached at the bottom. The WD (after 1950), WD-45, and WD-45 Diesels had the model designations on the radiator shell. The diesel had the diesel decal under the model designation on the side of the radiator.

No radiator shutters were used on tractor WD-45 number 146607 and after.

Air Filters

Prestreamlined WC and WF tractors used air cleaner assembly 226229, which had two instruction decals. The outlet to the carburetor was near the top of the filter canister. Streamlined WC and WF tractors had assembly 208191 with one instruction decal and an outlet to the carburetor near the bottom of the can.

The air filters used by WD and WD-45 tractors were of different construction. The WD-45 used a narrow center tube that extended from the air intake cap down to the oil cup. The WD air cleaners used a container within a container-type construction to funnel the air through the filter material.

Chassis Model WC Streamlined Dimensions, from the Advertising Literature (Form TL-164 421)

Dimensions—Length, 129 inches; width, 76 inches; height, 67 inches; ground clearance under rear axle, 29 inches; turning radius, 8 feet.

Standard Equipment—With rubber tires: belt pulley, rear wheel fenders, seat cushion, platform, starter and lights, radiator shutter; with steel wheels: belt pulley, rear wheel fenders, platform, starter and lights, radiator shutter.

Extra Equipment—With rubber tires: 150-pound drive wheel weights, PTO, power lift for implements, wide cushion seat and top assembly, special differential with 6 3/5-to-1-ratio for use with 7.5x36 or 9.00x36 tires; with steel wheels: PTO, power lift for implements, wide cushion seat and top assembly, 5-inch rear wheel extension rims, with ten 5-inch spade lugs per rim, special skid bands for front wheels, special skid bands for wide axle front wheels.

Starting Cranks

WCs with Serial Number 68610 and before used a one-piece starting crank, with a rotating handle at the crank end. WCs with Serial Number 68611 and after used a stub shaft that remained in the steering pedestal, and a removable starting crank. WFs used the two-piece crank from the start of production, as did the WD and WD-45. The starting crank was the same for all tractors.

The short crank was stored in short sheet metal clips. On WFs before Serial Number 1904, the crank was stored with one clip. On WFs with Serial Number 1904 and after, the crank was stored with two clips. On WFs, the crank was apparently held in the toolbox on tractors 68611-74329, when the same clips used on the later WFs were added. WDs and WD-45s used two clips as well, although the rear clip was slightly different from the WC and WF clips.

This is the WD and WD-45 gas clutch pedal and pad assembly (the WD-45 Diesel clutch pedal would have some bends down by the frame). The left toe plate, seen here just below the clutch pedal pad (the part the operator's foot goes on) was the same in WD, WD-45, and WD-45 Diesels. The thin tube at the bottom right-hand corner is one of Allis' high-pressure, low-volume hydraulic rams.

WC, WD, and WD-45 tractors with a single front wheel or the wide front axle were fitted with an extension on the engine end that also functioned as a U-joint to accommodate the different construction of the front pedestals for those tractors.

Transmissions

Transmission covers on WCs and WFs changed at the streamlined serial numbers. On both tractors, a reverse rod with a spoon trigger was used to unlock the reverse gear. Internally, tractors built before 1935 had the shift rods separate from the control cover, while on 1935 and later tractors, the rod supports are integral to the cover.

WD and WD-45 tractors used a "torque housing" to enclose the new transmission clutch. The transmission was actuated by the big hand lever curved at a 90-degree angle. The new clutch allowed the PTO to function as a live PTO—the transmission could be disengaged from the engine without disengaging the PTO drive.

WDs before Serial Number 127007 had a sliding gear transmission with a straight gearshift lever. The shifter knob was a round knob that was slightly flat. There were two different manufacturers of the knob, with one producing knobs with a slanted knurling around the edge, and one producing a straight up and down knurl with a slightly different stem. WDs serial 127007 and after had a constant-mesh transmission with curved gearshift lever and a shifter knob that was more spherical in shape. The knob was smaller in diameter than the previous knobs. Of course, there were extensive changes in the transmission involving gears and shafts. Also changed at the same time were the transmission case (which now incorporated a pedal stop stud) and the brake covers, which mounted on the transmission case.

The owner of this styled Model WF tractor has painted his caps black, perhaps an individual preference. The temperature gauge looks like a proper Allis gauge.

The wiring loom on this Model WD, here in black, is a modern replacement ribbed plastic housing that gives a much longer life than the original asphalt impregnated cloth wiring looms. The sediment bulb looks like a bulb from a kerosene tractor, since it has the extra nipple on the right side that was used for a starting gasoline tank.

The clean lines of the Allis design show well on this WD hood. In production, all caps would have been orange, with the temperature gauge masked off. Allis seems to have started using aluminized mufflers in the early 1950s.

Engine Clutch

Two different engine clutch housings were used on WC tractors: a standard housing and a housing for WCs with the adjustable wide front axle. The housings themselves changed over time, with a variety of different bolt patterns being used.

Starting with WD-45 engine 26993 (tractor Serial Number WD-169408), a new clutch lining was used that increased torque capacity. The clutch was again changed on engine WD-45 30956 (tractor number 172518) to add a pilot bearing. The Allis service manual recommended adding a pilot bearing to earlier tractors when repairing them.

WF tractors used a one-piece curved clutch pedal, while WCs used a two-piece assembly that allowed the operator to adjust the pedal to a comfortable distance from the seat. The bottom of the clutch pedal was notched to fit the clutch latch.

WDs and WD-45s used a different pedal adjustment method. The WD-45 Diesel clutch pedal shaft was shaped differently than that on the WD-45 gas. The clutch latch assembly was also different on the diesel. The clutch latch was notched to engage a nub on the clutch pedal.

WD tractors before serial 10505 used a flat sheet-metal cover for the belt pulley opening when the pulley was not fitted. Serial 10505 and after, including WD-45 family tractors, used a cup-shaped cover.

This WD taillight has the proper red lens, and is painted black, as the originals seem to have left Allis.

Differentials and Torque Tubes

There seems to have been a great deal of variation in WC differential and torque assemblies. Since the worm drive for the power lifts was driven from the intermediate shaft in the torque tube, a different torque tube with a shifter assembly was used in WCs with the live power lift (Serial Number WC-74330 and after). Standard WCs and WFs had a 4.47:1 rear-end gear ratio, while some Cane WCs and WCs with 36-inch rear tires used a 6.6:1 rear-end ratio (identified as "Hi-Traction" gears).

Hoods and Gas Tanks

There were a large number of different hoods used in this series of tractors. The unstreamlined WCs and WFs shared one hood, which was used on both low-octane and gasoline versions. The original hood was 204268, which was changed to 207885. The unstreamlined hood assemblies included the corner side sheets at the radiator end. Hood packing strip 202916 was used. Streamlined WC and WF tractors used the same hood sheets as the normal production WD, depending on fuel choice. The low-octane versions had holes to accommodate the small auxiliary gasoline starting tank, which was located under the hood on the right side of the tractor. Unstreamlined tractors used a main fuel tank that was squared off at the ends, while

The WD-45 had its toolbox located in the same place as the WDs, on the left fender. The WD and WD-45 tractors that had starting cranks had them mounted on the inside of the right fender.

Snap Coupler latch assemblies changed at least four times in production of the WD-45 tractors. The latch assemblies went through quite a bit of stress, and were also liable to be damaged by backing into something. The arms on this tractor are replacements.

From the front, the WD doesn't look all that different from the WC. Grills were painted orange. The toolbox location on the inside of the left fender is correct for the WD.

This WC has the cast-iron rear end. Allis also used a pressed steel rear end, due to the difficulties of obtaining cast-iron rear ends during World War II. Tractors with the pressed steel rear end received a suffix letter on the chassis serial numbers to denote the difference.

The seats on the WD tractors changed at Serial Number 127007. The long seat lugs in the bar assembly can be seen at the top of the photo. The mounting plate at the bottom of the seat is also different. The auxiliary light socket, which does not have its wiring attached, is located next to the hydraulic line.

the streamlined tractors used a tank with a teardrop shape at the rear portion.

WD and WD-45 models without power steering shared hoods—the 220616 for gasoline engines and 223563 for low-octane fuel units. WD-45s with power steering had two hoods as well, for gas and low-octane fuels.

WD-45 Diesels, with their longer six-cylinder engines, had two hoods available—one for tractors with power steering and one for those without.

Instrument Panel

Rather than having the gauges spread all over the tractor, the WD-45 Diesel had an instrument panel on top of the hood that grouped together the water temperature, traction booster, and oil pressure gauge.

Gas Caps

Tractors before WC-511 used a hinged gas cap. Some tractors apparently used the winged cap (207244). From tractors WC-511 to WC-2044, a cap with a chain retainer was used. Tractors WC-2045 and after used cap 208065.

Sediment Bulbs

WCs before WC-547 used an Imperial fuel filter assembly. On WCs 547 and after, to the end of unstreamlined production, and on all unstreamlined WFs, an A-C filter assembly was used. Both glass and metal bowls were available.

Streamlined WC and WF tractors, WC-92330 through WC-107704, and WF-1904-2303, used a fuel filter with a screw-on sediment bulb instead of a metal bail to hold on the bulb. The screw-on bulb was glass. For tractors with low-octane fuel, sediment bulbs with two inlets instead of one were used. Other tractors had metal and glass bulbs optional. WDs used the same fuel filter on both low-octane and gasoline tractors, but a plug was used on the gas tractors to close off the extra inlet. WD and WD-45 tractors had metal and glass bowls.

Fenders

Unstreamlined WCs used squared-off, arched fenders before serial 74330. The toolbox was located on the inside of the right fender, near the rear, and was hinged at the back lengthwise along the box. Left-hand and right-hand foot plates were bolted to the inside of the fenders just ahead of the axle openings. Streamlined tractors serial 74330 and after used stamped clamshell fenders with foot plates. The toolbox was bolted to the inside of the frame channel and was hinged on the side. The later toolbox was wider and flatter than the earlier box.

Unstreamlined WF fenders were square in section and more of a teardrop shape lengthwise. The fenders also had side panels to form the forward part of the platform enclosure. These fenders did not have an enclosed toolbox, but rather a tool tray bolted through the fender and side panel.

The operator's seat and controls on WD and WD-45 Series tractors are offset slightly to the right, as can be seen here on this WD-45 Diesel. The goal was to give the operator a better view of the front wheels and cultivators.

The WD replaced the mechanical lift of the WC with a high pressure hydraulic system, and had an extensively revised driveline behind the clutch that included a new clutch housing. The battery box was relocated to under the fuel tank, and the shifter was placed much closer to the operator.

The hydraulic lift rocker and lift arms are visible on this 1951 WD. The rear axle casing is considerably more substantial than the one on the WC. WDs had a "platform" on the back of the tractor (visible under the rocker arms) that acted as a handy step to mount the tractor, and also as a PTO shield.

Streamlined WF fenders were the clamshell type, but were longer and squarer lengthwise than the WC fenders. The same toolbox as on the WC was used, bolted inside the left fender. The tractors used fender supports that were flat-shaped plates.

Drawbars

Several different drawbars were used on WC tractors. The original drawbar was a straight, one-piece, nonswinging drawbar. If the WC was equipped with a PTO before 1944, a non-ASAE standard drawbar was used. This drawbar had two pieces bolted together that dropped lower to the ground, to clear the PTO shaft.

After 1944, ASAE standard drawbars that could be used with non-Allis equipment were used. A fixed drawbar was standard, but a swinging drawbar was also available. All WC drawbars were bolted at the rear to a support that connected with the rear platform, which was considered part of the assembly.

WF tractors had as standard a swinging drawbar. If a PTO was fitted to the tractor, a nonswinging fixed drawbar was used.

Front Axles

WCs had available the standard dual-wheel narrow front axle, the single-wheel front axle, the adjustable wide front axle and the higher arched cane axles. The WD and WD-45 also had the same axle options, with slight differences in the axle support plates on the wide front axles. On WD (including WD-45s) single-front-wheel tractors, the fork assembly before Serial Number 25129 was one piece, while on 25129 and after the fork was separate from the fork steering shaft. The front supports on WD and WD-45 tractors also varied by the axle type.

Rear Axles and Housing

The WC tractors used two different rear axle housing types. Both housings used dropped final drives to gain crop clearance. The main axle housing was cast malleable iron. However, for several years during World War II, pressed steel housings and bolted-on brake housings were used. These tractors are identified with an "S" in the tractor serial number. The malleable housings received a cover housing at serial 67328 and after. The malleable housing had the serial number pad cast into it on the rear of the differential area.

WCs had two different types of pinion shafts, which transmitted power from the differential out to the final drives. Tractors before 86405 had a removable pinion, while tractors with serial 86405 and after used an integral pinion on the shaft. There are small differences in the outside casings.

WFs used a drop housing type of rear axle that was of different construction than those on the WCs, although many internal parts were common between both tractors.

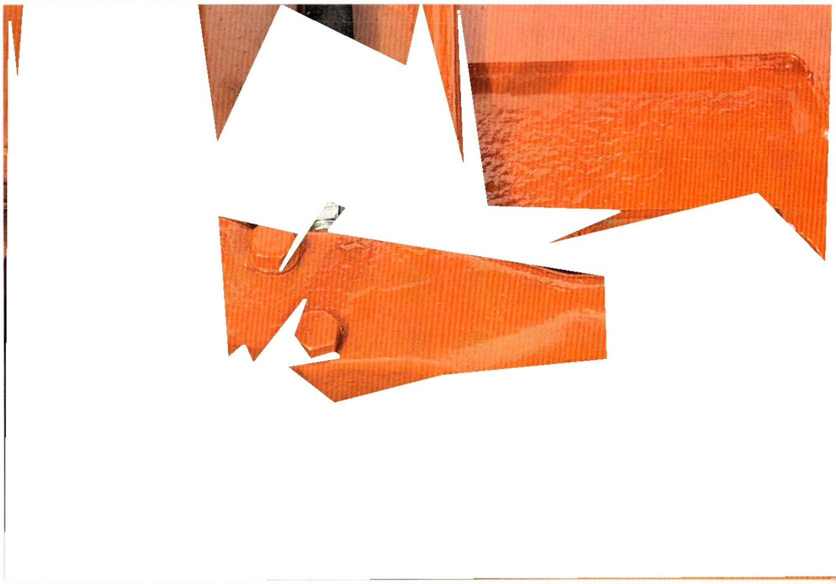

This is the WD's hydraulic hold positioning valve. The hydraulic pump is under the valve to the right. Hoses were probably painted orange in production. The knobs seen on the hydraulic pump are what those big hydraulic pump instruction decals refer to!

This is the location of the WD-45's hydraulic hold positioning valve, on the right side of the tractor just forward of the battery box. The hydraulic pump is under the valve. Everything's orange.

The WD-45 Diesel's hydraulic pump instruction decal is seen here on the battery box in the correct position. The round shape in the frame rail underneath the starter is the belt pulley tube opening cover, used when the tractor was shipped without the belt pulley unit. This is the cover used after WD tractor 10595, which of course would include all WD-45 series tractors.

Seats

WDs before serial 127007 had a seat bar assembly that had two "ears" that attached to the shock absorber. On WD and WD-45 tractors after this serial number, a bar with adjustable slotted ears was used.

Steering

WFs used a draglink system with the actual gearing located on the operator's platform. Tractors previous to Serial Number 1904 used a Lavine steering gear, while Serial Number 1904 and after used a Ross type of steering gear. Both used steering wheel 205623, but the Ross gear used a "worm shaft stopper," 206882, that covered the hole in the middle of the nut that held the steering wheel on.

WCs, WDs, and WD-45s used a long sloping shaft from the steering wheel to a U-joint, then another shaft to the front support, which enclosed the actual gears. WCs originally used steering wheel 202260, which was superseded by 229677. The wheel and shaft assembly was supported by a support assembly. U-2116 was originally used, but this was superseded by 208869. A brace from the support assembly to the differential case was added at tractor WC-29809.

The front supports (also known as the pedestal) on the WC, WD, and WD-45 tractors were different, depending on what axle was fitted. The front support for the single-front-wheel and adjustable wide front axle were the same, while the support for the standard dual-wheel narrow front was separate. The WC front support for the single-wheel and adjustable-axle models was originally part U-3946, but notes in the parts book say to replace the older axle with the WD support for service purposes. Likewise, the dual narrow support was replaced.

The worm gears located in the supports had a standard gear ratio (the worm was part 201111) but a higher turning ratio worm (220662) and gear (220675) were also available for later replacement.

Two slightly different steering wheels were used by WD and WD-45 tractors. The wheels were originally painted orange, but the paint was changed to black, probably in January 1955.

WD-45 Power Steering

WD-45s fitted with Barnes power steering pumps had nameplates on them. The pump for the WD-45 had an oval-shaped nameplate (Allis part number 227335, Barnes part number PA-8471), while the WD-45 had a rectangular-shaped plate with rounded corners (Allis part number 227535, Barnes part number GC-1498). Both plates were marked "John S. Barnes Corp, Rockford Illinois," and had patent number 2,051,414. The plates, which were made of aluminum, were also marked "Patent Pending."

WD-45s with power steering and dual narrow front or single front wheels had a mud guard assembly to protect the power steering.

The WC and WF were fitted with hand brakes. The levers extended from the brake housings up to each side of the operator. The levers varied. WCs with the pressed steel rear ends had different levers than those with cast rear ends. WFs had their own brake levers. WCs with 7.50x36 or 9x36 rear tires, but without fenders, had different brake covers from the other WCs.

WD and WD-45 used foot brakes. Tractors before WD-24801 without hydraulic cylinders used a brake pedal stop. This was built into the right-hand toe plate on Serial Number 24801 and after. WD-45s serial 177712 and after received a new brake lining with improved life.

Power Lifts

The WC tractors were available with either a heavy-duty or light-duty power lift, both driven by worm gears. The operator stepped on a pedal to put the lift into motion. Both lift units used pedal 202716, except for heavy-duty lifts with lister attachments, which used pedal 210690. A pedal pad assembly, 203498, was used on tractors built before 1935.

Dimensions for Model WD from the Advertising Literature

Overall Dimensions—Wheelbase, 88 inches; overall length, 128 inches; overall width, (narrowest wheel tread) 74 inches, (widest wheel tread) 102 inches; overall height, (over muffler) 81 1/2 inches; ground clearance under final drive housing, 28 1/8 inches; shipping weight fully equipped, approximately 4,000 pounds.

Extra Equipment—12x28 tires (substitute), belt pulley, remote hydraulic ram, front wheel weights, rear wheel weights.

Hydraulics

The WD and WD-45 had hydraulic lifts. A hydraulic ram acted on a rear-mounted rocker shaft with two arms. The WD-45 also had a transport valve that worked with remote implements. The Allis hydraulic systems ran, for their day, at very high pressures.

Allis-Chalmers recommended that when using non-Allis hydraulic rams, the capacity of the rams should not exceed 40 cubic inches. Non-Allis rams were intended for lower-pressure hydraulic systems. The higher-pressure Allis system didn't require as much ram volume. Using rams of over 40-cubic-inch volume could result in the Allis hydraulic system running out of fluid. The WD-45 had an additional round hydraulic reservoir tank. The hydraulic fluid also lubricated the belt pulley bearings and gears, meaning that damage could result if the system ran out of fluid. A magnetic plug was located at the bottom of the housing to capture metal particles, which could damage the system.

PTOs

Two different types of PTOs were used on WC tractors. The first type was driven from the belt pulley shaft, with a long PTO shaft that angled downward from the drive area under the rear axle to a support attached to the drawbar. This PTO shaft assembly was used before 1935. This unit did not have a shifter.

The later PTO assembly used on 1935 and after WCs and all WFs was driven from the transmission. The unit was driven by gears dropping under the transmission to a housing. The PTO shaft ran from the housing back. A shifter rod was located on the right side of the unit. A front-mounted mower could be run directly from the housing, instead of having to run a shaft from the rear.

A sheet-metal PTO guard with a cast bracket was used until approximately the start of 1940, and after that time an all-sheet-metal guard with braces (and warning decal) went over the top of the end of the shaft.

Belt Pulleys

On WDs and WD-45s equipped with PTOs from the factory or dealership, an arrow was stamped on both the tractor frame and the belt pulley housing. When the belt pulley is engaged, the arrows should line up. Some Allis operators manuals say that this arrow was stamped by the dealer, while others do not mention who stamped the arrow. The arrows indicated the relative locations of the housing and the frame to determine the amount of backlash.

Wheel Weights and Front Support Weights

At least three different wheel weights were available for WC and WF tractors, depending on the type of wheels used. Weight 203594 was used with spoke wheel 202069 or disc wheel 207509 (11.15x24 air tires). Weight 203592 could be used as an outer weight with 203594, and also was used with cast wheel 208916 (11x28 air tires). Weight 203998 was used with cast wheels 203997 and 204010 (used with 9.00x28, 7.50x36, or 9.00x28 air tires).

WD and WD-45 tractors used a rear wheel weight that could be used as an inner or outer weight. A disc-shaped front wheel weight was also available.

WD-45s had front support weights available. These weights were curved to fit over each other, and were stacked on top of the front support ahead of the radiator.

This B was manufactured in 1940. The classic lines of one of the most influential tractors of all time show well—small size, excellent visibility from the operator's position, and simplicity were the features of a design that sent every competitor back to the drawing board and kept them there for the next 20 years. This tractor has the high arched nonadjustable front axle.

CHAPTER 7

Models B, C, CA, IB, RC

The Allis-Chalmers B, when it first emerged in 1937, stunned the tractor industry. Here was a tractor that was very small, very light, and very cheap, but still a very capable row crop tractor that could put in a full day's work. The novel torque tube construction combined with attractive sheet metal, manufactured in large part by outside companies, produced a tractor that was a full generation ahead of anything else in production. Other companies quickly scrambled to build a small tractor to compete, but Allis' broad patents on the torque tube design ensured that competitors either produced a poor tractor or paid Allis some hefty patent royalties. The design of the B, originally intended for single-row cultivation, was changed slightly to produce the companion two-row C. The C was then later changed with some driveline modifications (an extra clutch to provide a live PTO) into the CA.

The first Bs were sold under catalog number 290143. These are referred to as the '37 Bs, or as the first 97 Bs. These were preproduction tractors with many handmade parts, including flat-sided fenders and a unique front axle.

Regular production Bs were produced with a high number of parts wholly or partially manufactured by outside companies. This first generation of Bs did not have electrical equipment. Allis called the Bs with electric starting and lighting "Deluxe Bs." They appear to have entered production with Serial Number 43220.

The IB was an industrial version of the Allis B, reinforced and reduced in size. The Frank G. Hough Company was apparently responsible for the first conversions, and supplied most of the special IB parts until 1945. It is unknown whether Hough handled assembly of all IBs until that point, or whether Allis purchased the parts from Hough and assembled the tractors itself. The IB featured a shorter torque tube, a battery box located in the operator's platform, a square extra frame around the tractor, and a different front axle and a rear axle that was rotated 90 degrees to lower the tractor.

The Allis-Chalmers C joined the B in production in 1940. It was essentially a B redesigned with a narrow front wheel to cultivate two rows at a time. The C was available with several front axle configurations. The C was replaced by the CA in 1950.

CAs were fitted with many of improvements. The most noticeable was the arrangement of another clutch located on the rear axle (which disengaged the power to the rear wheels, while leaving the power to the transmission unaffected) to provide a live PTO, similar to the WD tractor. Traction Booster hydraulics were also fitted.

Variations

B Asparagus tractors were high clearance tractors, with 8x38 rear tires and 4.00x15 front tires. The asparagus tractors were modified over time to match the regular B production including the addition of electrical equipment with the Deluxe B, triple rib front tires, and other parts. B Asparagus tractors had a front axle tread width of 45 1/4 inches.

B Orchards were produced in apparently tiny numbers, with catalog numbers being assigned in both April 1940 (Allis catalog 290336) and the early 1950s. These tractors apparently used some IB parts, but without the outside IB frame and with the longer torque tube of the regular B.

Allis sold some Bs as cane tractors (with 5.50x16 front tires and 9.00x36 rear tires) and sold some Bs to

LEFT
An archival photograph of a late B shows the model designation sticker up on the grill, orange front wheels and rims, but orange wheels and silver rims on the rears. The steering wheel looks all orange, and the larger muffler and a black generator are being used.

NEXT PAGE TOP
Some unusual points to note on this B: The oil filter doesn't have the decals on it, the magneto and spark plug wires are completely painted, the padded seat back and seat pad seem to be different colors, the front rims have no stripes, and the rear rims seem to have been painted black or some dark color. The caps are all painted orange, and the cap and bolts to support the gasoline starting tank are seen just above the "Allis-Chalmers" decal. *State Historical Society of Wisconsin*

NEXT PAGE BOTTOM
This B is about 10 years too old to have the model designation sticker, but it does answer the question, "What is that thing?" at shows. The shifter boot is orange, as was the original, but the boot itself is a more modern replacement—Allis had a water leakage problem with the older boots.

These photographs were taken by International Harvester as part of a patent dispute with Allis-Chalmers. This was a standard production B that was sold by Allis to IH for comparison purposes. The info tag on the manifold is visible. The trade name decal appears to be something other than the regular orange paint. The "squashed tube" muffler is poking through the hood. *State Historical Society of Wisconsin*

ALLIS-CHALMERS

ALLIS-CHALMERS
B

Someone working on the assembly line must have stepped on this Model B's drawbar while the paint was still wet, as it has step marks on it. The rear rims have had something dripped on them. The trade name decal has the bolt for the gasoline starting tank through the dash in "Allis-Chalmers." *State Historical Society of Wisconsin*

Most IBs, if not all, had a reinforcing frame around them to keep the torque tube and engine block from fracturing from sudden shock or high loads. Some early IBs, with B serial numbers, may not have had the frames. The unique IB front axle assembly can also be seen here.

Much rarer than the regular B is the IB Industrial tractor. Intended for use as a tug in factories, the IB Industrial has a lower height. The final gear housings were rotated 90 degrees to lower the back, while the front had a different axle.

Thompson. Cs were also sold to Thompson. Part 212526 is a "Final drive housing (left hand) for High Clearance Model C Tractor." Apparently, final drives were different, along with tires and front axles.

After the C went out of production, CA Canes were built for Thompson.

Engines

The original Bs (before Serial Number 101), known as the 1937 Bs, used a Waukesha 3x4 engine. These tractors had a Zenith carburetor. B and IB tractors from B-101 to B-64501 used an Allis-Chalmers designed 3 1/4 bore engine that had a "BE" prefix on

The rear end of the IB, because of the different height of the tractor, as well as the different uses, was quite a bit different from the B farm tractor. IBs had a different taillight from other Allis tractors. The shifter boot seen in the back is the correct shifter boot for the B and IB tractors from this time and before. The hitch side support plates and back plate are regular production IB parts, but the parts books show an industrial pintle hitch assembly with a rope to attach the latch.

the engine serial number. (An extensive number of parts on this engine were machined and assembled by vendors.) B and IB tractors after these serial numbers, and all C and CA tractors, used an Allis-Chalmers–designed engine with a 3 3/8-inch bore and a serial number with a "CE" prefix. Some power unit engines based on the same engine had "R" and "CR" prefix serial numbers, while other power units had the same prefixes as the tractor engines.

Early B tractors did not have electric start. Later Bs came in two versions, the standard and "Deluxe" tractors, which had flywheels with ring gears for starters.

When the oil pump bypass is set correctly (12–15 pounds of pressure), the needle on the oil gauge should register in the "ORM" letters of "NORMAL" on the gauge.

On older B and C engines, both timing marks are straight lines. On newer engines, the crankshaft gear has a straight mark, while the magneto gear has an oval mark. On the CA engine, the timing mark on the distributor drive is only marked with an oval if the tractor was originally fitted with a magneto. The camshaft gear also has an oval. If the tractor had a battery ignition, the marks are straight.

The C was the excellent answer to the need for a two-row tractor in the B size. With many components that were used in B production, the C proved to be an extremely popular tractor that continued on into the CA, D-10, and D-12 families.

This C, from 1950, is about 1,700 tractors away from the last C produced. A careful examination of this very interesting tractor will see some points that are far from your average production C. For instance, the extra gearshift. This tractor was fitted with an auxiliary transmission, that Allis-Chalmers experimented with.

The B, C, and CA engines had two different manifolds, depending on fuel. The gasoline manifold had a square cover, while the two-fuel manifold was a longer rectangle with the two lower corners beveled off. Both manifolds gave engine timing and tune-up information.

After engine CE190257, the oil intake pipe and screen were changed to the floating type. Starting with

The fenders on this C have been lengthened to accommodate another gearbox that added another range of gears to this tractor. While the precise details of this tractor are not known, it is known that a company produced a small gearbox that went between the torque tube and the gear box of the tractor. The company offered the gearbox to Allis; it isn't known if Allis instigated the project. This tractor could be the gearbox manufacturer's test tractor, the Allis test tractor, or a tractor converted for a customer, such as a farmer.

With the C, the two rows of crop moved down either side of the front spindle, then passed just inside of the drop final gear housings when cultivating (of course, when the drawbar was removed). It was a brilliant answer to the question of how to create a two-row tractor from a one-row—one wonders whether IH copied Allis, or Allis stole an idea back from IH!

CA tractor 30574, engine CE 18419, the rocker arms were made of cast iron instead of steel stampings. On engine CE-191558, Allis used the intake rocker arms on both the intake and exhaust arms to control oil. On engine CE-161057 and after, the exhaust valves received rotators.

B Engine Serial Number Suffixes for 3 1/4 -inch Bore Engine

Allis provided engines with various compression ratios to ensure that full engine power was available to all users, including those who farmed at elevations where low air densities adversely affected horsepower.

Gasoline

G=Under 3,000 feet altitude

B=3,000–6,000 feet altitude with two cylinder head gaskets

B=Over 6,000 feet altitude with one cylinder head gasket

Distillate engines

K=Under 3,000 feet altitude

G=3,000–6,000 feet altitude

B=Over 6,000 feet altitude with two cylinder head gaskets

The extra gearbox on this C and the revisions in the platform to accommodate it are clearly seen in this photo. If this is the same gearbox used in Allis testing, its downfall was partially the small size of the box—the small amount of oil inside quickly got too hot.

This Model C was an all-fuel, which could burn kerosene or distillate. The mountings and filler cap for the small gasoline starting tank used with those tractors can be seen on the hood just above the "Allis" on the trade name decal. Of course, the oil filter is a more modern replacement.

This is an archival photograph of a CA. Notice the position of the "CA" decal, the already rusty muffler (it was not aluminized), and the orange starter, steering wheel, and distributor. The rear rim is silver. *Photo courtesy of Norm Swinford*

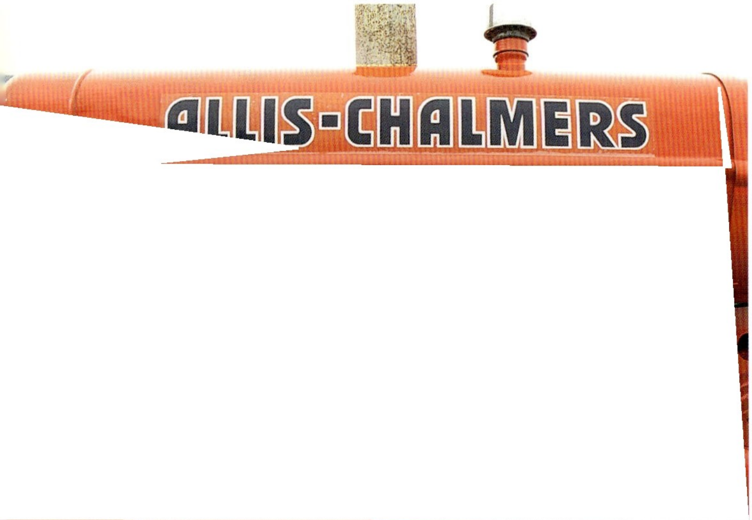

The specification plate is seen here on the Model B's manifold. The air cleaner decal is the Donaldson decal found on Allis tractors of this time period. The air cleaner cap, of course, would not have been chromed in production!

On this nonkerosene Model B, the trade name decal is placed off the end of the sheet-metal "bead," just under and to the right of the last "S" in "Allis-Chalmers." The oil filter and spark plug wires are modern replacements.

The Allis-Chalmers CA is a remarkable tractor for its size and era—a small tractor with a decent amount of power, with a live PTO that was useful for pulling Roto-Balers and other PTO-driven machines, and good hydraulics—it is still in demand today for belly mowers and other uses.

A close-up of the left side of the IB engine shows the engine specification plate, the generator with the unpainted data plate, and the air filter with Donaldson decal. This generator is painted orange. At some point in the late 1940s or 1950s, Allis may have changed its painting practice from not painting the generators, which came from their factory painted black, to painting the generators with the rest of the tractor and just masking off the data plate.

RC Engine Serial Number Suffixes for 3 3/8-inch Bore Engine

Gasoline

G=Under 3,000 feet altitude (5.0:1 compression ratio)

B=3,000–6,000 feet altitude with two cylinder head gaskets

B=Over 6,000 feet altitude with one cylinder head gasket (6.0:1 CR)

Distillate engines

K=Under 3,000 feet (4.7:1 compression ratio)

G=3,000 to 6,000 feet altitude (5.0:1 CR)

B= Over 6,000 feet altitude with two cylinder head gaskets (5.7:1 CR)

CA Engine Serial Number Suffixes

K=Kerosene

P=Gasoline

Carburetors

Several different Zenith and Marvel Schebler carburetors were used, depending on age and fuel type. Alcohol carburetors were available on B, C, and IB tractors, as well as low-octane and gasoline versions.

All but the earliest CAs should have a model designation decal. Headlights seem to have always been black, but generators and starters seen on most CAs in sales literature are actually orange, suggesting that Allis made a change in the late 1940s or early 1950s.

The carburetor on this Model C has been painted black rather than the original orange, but the rest of this area looks fairly original. The fuel line and air cleaner hose would also have been painted orange.

Cooling System

On engines before BE-23815, a nonpressurized radiator was used with a tall water manifold (referred to as an "elbow"). Two water manifold castings were used, 207464 for gasoline-only tractors, and 208015 for low-octane tractors. A nonpressurized water temperature gauge was used, originally 208684, which was superseded by 213675. Tractors without temperature gauges had threaded plugs to fill the resulting hole in the block. No thermostat was used on these tractors.

Tractors with engines numbered 23815 and after used a flatter thermostat housing, which replaced the elbow-shaped water manifold used previously. A pressurized radiator was used with these tractors. Two thermostats were used, one for each fuel option. A pressurized gauge (originally 208684, superseded by 213675) was used. Tractors without a temperature gauge were fitted with a plug.

The engines beginning on tractor B 126210, IB 3238, and CR-60157 and after used a bypass-type water manifold that eliminated the need for a radiator shutter. This design also used water temperature gauge 213675.

The B, C, and CA pressurized radiator and radiator cap were designed for 7 pounds of pressure. The CA had two different thermostats, both with two different temperature settings. The Dole thermostat had to be used with two gaskets to provide clearance between the valve and the thermostat housing, while the Bishop and Babcock did not require gaskets. The gasoline thermostats opened at 135 degrees Fahrenheit, while the low-octane fuel thermostat opened at 165 degrees. Engines CE-195136 and up used a water manifold with a built-in bypass.

The Bs before 101 did not use a radiator curtain. The radiator grill had a reinforcing rib at the bottom of the shell. On Bs and IBs after serial 101, radiator curtain 208104 was used before Serial Number 208014, when it was replaced. Radiator curtain 209540 was used on B tractors between 22260 and 43301, and on IBs before B-55301. All radiator curtains were made of olive drab cotton duck fabric. B and IB tractors were then fitted with radiator shutters.

B after serial 101 used a different grill and shell than the previous tractors, having a curvier shape. IBs had their own grill and shell assembly. C tractors had yet another grill that did not reach back as far under the tractor as the B and IB grill. Cs used a curtain assembly, with no curtains being listed.

Exhaust System

The flattened-pipe style exhaust was used on gasoline B tractors with engines before Serial Number BE-43896, and on gasoline IBs with tractor serial numbers before B-55301. The oval can underhood–type muffler was used on B and C tractors with engine serial numbers between BE-43896 and B-114400, and on IB tractors from serial B-55301 through IB-2397. Tractors after these serial numbers used the tall, narrow can muffler with a flap for mounting to the manifold. This last type of muffler went through a hole in the hood.

Low-octane B and IB tractors used the flattened pipe exhaust on B tractors with engine serial numbers before BE-43896 and IB tractors before Serial Number IB-1001. B, C, and IB tractors after these serial numbers used the underhood, oval-shaped can-type muffler till the end of production, never switching to the upright muffler.

Mufflers on CAs differ by type of fuel used. The low-octane fuel tractors used the large oval underhood-type muffler. The gasoline version used the tall muffler, with a flap that bolted to the exhaust riser of the manifold. Later flap-type tall mufflers were aluminized.

Sediment Bulbs

Both metal and glass bowls were used. On Bs before serial 101, sediment bulbs consisted of a filter assembly and sediment bowl that screwed together, combined with a bail assembly, rather than the traditional bail assembly that supported the bowl. On tractors after 101, two different sediment bulbs were available, one with a screw-in bowl without a bail, and a more traditional bowl with a bail.

Air Cleaners

Bs before serial 101 used an air cleaner with a clamping strap and a long air tube. The air outlet was near the top of the canister.

Later B, IB, and C tractors' air cleaners had a short tube going to the carburetor. Bs and Cs used a stack brace. Two decals were used—one on the main can that gave the cleaning instructions, and an oil level decal on the oil cup that was discontinued in 1948.

Two different types of air cleaners were used on CAs. Tractors with engine numbers before CE-13732 used assembly 207479 (which had a tapered bottom cup), while engine serial number CE-13732 and after used assembly 225856 (Donaldson part FCA04-6000), which had a different mounting and a straight bottom cup.

Governor Controls

The governor control (commonly but incorrectly called the throttle) on Allis Bs before serial 101 used a cast quadrant, which was also used on WCs and early WFs. These quadrants had the words "low" and "high" on them. Bs after 101 and all Cs had a simpler quadrant. IB tractors had a combination hand lever and foot pedal. IBs before 1001 used a different assembly than later IBs.

Governors used with magneto ignitions were different from governors used with battery ignitions. The battery ignition unit had a distributor gear drive located on the shaft assembly, whereas the magneto-type governor did not. The battery governor had a combination shaft and governor gear assembly, while the magneto governor had a shaft separate from the gear.

Transmissions

B tractors before serial 101 used a shift lever with an integral knob. A flat metal dust shield was used. Tractors after serial 101 used a rubber dust shield. Tractors 101 to 124632 (including IBs and Cs) used dust cover 226863, while tractors 124638 and after used dust cover 231454, which was designed to elim-

Model B starting cranks seem to end up sticking out of the steering shaft hole like this, or out of the toolbox, which is covered by the flap seen just below the crank. Early nonelectric Bs have a smooth torque tube without a hole and flange for the starter. Later Bs without electric starting would have a torque tube with the hole and flange, but would have a sheet-metal cover. There actually should be a clip to hold the starter crank just below the steering shaft hole.

Chassis Model B Dimensions, from the Advertising Literature (Forms TL-283, TL 329 (1-41))

Overall Dimensions—Wheelbase, 73 inches; length, 111 inches; width, 53 inches (44 tread); height, 62 inches; turning radius, 7 3/4 feet; ground clearance, 22 inches over rows.

Regular Equipment—Rubber tires and inner tubes, liquid ballast in all tires, rear wheel fenders, cushion seat. (January 1941)—Rubber tires, fenders, thermostat, radiator curtain, cushioned seat.

Special Equipment—Belt pulley and PTO assembly; distillate burning equipment; oversize 9-24 rear tires; 13-pound rear wheel weights; adjustable front wheel tread equipment, (narrow style) 38-, 43-, and 48-inch treads, (wide style) adjustments to secure 50-, 55-, and 60-inch treads, factory or field; wide adjustable rear wheel spacings of 48 52, 56, 60 inches, factory or field. (Special Equipment, January 1941) Starter; lights; radiator shutter; oversize 9x24 rear tires; rear wheel weights; adjustable front axle; swinging drawbar; side hill hitch for swinging drawbar; hydraulic lift.

A top view of the Allis B. The chrome caps were, of course, painted orange. The muffler is a modern replacement, which is quieter than the original and does a better job of keeping exhaust fumes out of your face during a day of cultivating. The one orange cap seen here is the cylinder head breather cap.

Operating controls on a B are pretty simple. The steering wheel here is all black—Allis changed the specifications for its steering wheel colors at least twice during the production run of the B. The shifter boot is a more modern Allis replacement that has fewer water leakage problems. Water leakage into transmission cases can cause rapid corrosion, not to mention possible fracture of the transmission case in freezing weather. The original Allis boot did not have the multiple "baffles" seen here, and it used a different method to seal the boot against the shaft and the case.

inate a problem with water leakage into the transmission casing. Tractor B-120982 and after (including Cs) and IB-2600 and after added a rubber shift knob on top of the integral knob.

CAs received rubber shifter knobs at tractor CA-18971 and after. A dust cover was used on the shift lever. At tractor CA-30611, the foot-clutch pedal changed from 226308 to 210866. The final drive clutch was actuated by a hand lever that had a rubber knob on top.

All tractors with lighting systems used two headlights and one taillight. The taillight changed at serial B-113894. The first taillight lens was clear glass, in a can-shaped shell. The later taillight lenses were ruby glass in a teardrop-shaped shell. Both were made by Guide Lamp. IBs used their own taillights, which had more of a diamond shape when looking at the side.

Clutch Pedal

Before tractor B-3481, a clutch pedal assembly comprising three parts welded together was used. B-3481 and after (including IB and C) used a one-piece cast pedal. Clutch pedal 210184 was used on tractors without electric starting, while two different clutch pedals were used on tractors with electric starting: 210866 was used on tractors B-43220 to B-124037, and IB tractor B-55301 through IB-2836. Clutch pedal 226308 was used on B-124037 and after and IB-2836 and after.

Hood/Sheet Metal/Fenders

Hoods varied by serial number and fuel type, with different holes for auxiliary fuel tanks and for vertical exhausts. Bs before 101 had their own hood that was handmade, not stamped. Hood straps varied with electric starting.

A variety of fenders were used on Allis' small tractors. The Bs before Serial Number B-101 used a flat fender assembly made from several pieces. After serial 101, fenders changed several times on the standard Bs. Bs fitted originally with steel wheels used different, fuller fenders than the air tire tractors. Narrow 36-inch tread Bs and asparagus tractors used a flat sheet-metal fender and narrow supports.

C tractors used different fenders than the B or IB. The fenders on tractors before C-109 had a small piece

A top view of the IB shows orange caps, black headlights, a nice Allis temperature gauge, and an aftermarket exhaust cap. The exhaust cap is one of those useful things that small manufacturers and farmers found out about before the tractor manufacturing companies discovered them.

The seat shock on this Model C was from Monroe. The Monroe "easy rider" seat was a popular aftermarket seat that provided greater comfort and shock protection, and made the tractors, which originally had bench seats, a little easier to get on from the ground. The light is also probably aftermarket, and was probably intended to illuminate an implement at night, rather than to act as a red taillight like Allis original equipment.

spot welded onto the nose that dropped below the rest of the fender and had no stamped reinforcing ribs. Regular production fenders had a flat bottom with no nosepiece, and reinforcing ribs stamped into the fender.

IB tractors used fender assemblies 210436/219908, but had two support plates running on the inside of each fender instead of the angle iron supports of the B and C. These fenders were also used on some WC tractors with 9.00 tires.

Unlike the other tractors, the CA used a sheet-metal operator's platform, much like the larger wheatland tractors of other manufacturers of that era. CAs had two optional hood sheets, one for gasoline tractors (with one hole for fuel tank access) and one for low-octane tractors (with two holes).

Fuel Tanks and Shroud/Toolboxes

Bs before serial 101 used a fuel tank that sat on the shroud/toolbox assembly, cushioned by two pieces of rubber beading. No auxiliary fuel tank was used.

After serial 101, a strap was added to the rear of the fuel tank that attached to a bracket on the shroud/toolbox. Two pads were used to cushion the tank.

The clip for the starter for hand-crank tractors was located on the back of the shroud. This clip was not fitted on tractors with electric starting. However, many operators obtained hand cranks for their electric start tractors, sticking it in the hole of the toolbox.

Small auxiliary fuel tanks for starting gasoline tractors were fitted under the hood for low-octane fuel tractors. The tanks had their own caps and a shut-off valve before the tank.

B Cabs

Canvas cabs were released by Allis for B tractors in 1939. The cabs were made by Bostrom.

Electrical Systems

Magnetos were used on tractors with engine serial numbers before CE-149840. The battery ignitions replaced the magnetos, although magnetos may have been available afterward as an option. Both systems used one spark plug wire retainer. The magneto used a kill switch. The battery ignition used a turnkey that activated the coil. A pull rod was used on both versions to activate the starter. A terminal cover was used where the battery cable bolted to the starter.

A Fairbanks Morse magneto was used on tractors before B-123996 and IB-2836. Fairbanks Morse battery ignition units strongly resembling magnetos were used on engines BE-43946 and later, and on CE engines. Up through BE-46895 and CE engines before CE-101, an ignition unit with a single battery contact (Fairbanks Morse model FM-T4B) was used. After this, an ignition unit with two battery lugs was used. On tractors with Serial Number B-123996 and after, and IB-2836 and after, an upright distributor was used, made by Delco-Remy.

Spark plugs were Allis-Chalmers 212056 for gasoline and U-2980 for distillate.

On CAs, two different pieces were used as the front half of the instrument box, one for magneto ignitions and one for the distributor ignition (210913, 225309). Several different light knobs were used. One taillight was used that was teardrop shaped and had a ruby red glass lens. Two headlights were used, with the lamp assemblies coming in a right-side and a left-side version. On CA Serial Number 32446 and after, a socket for the auxiliary warning lamp was fitted, although this socket could be fitted to older tractors as well. The socket was produced by Aetna Automotive Products.

The CA has several distinctions setting it apart from the other Allis small tractors. The sheet-metal platform for the operator's position is visible. The long lever to the right of the seat was the axle clutch lever, effectively yielding a live PTO. To the rear is the hydraulics control lever, which should have an instruction decal near it.

The Model CA's taillight has the red lens. The shift knob was probably painted orange at the factory. The steering wheel has a "suicide" knob on it. Some Allis tractors actually had these when they left the factory, but they had been eliminated by the days of the CA. However, probably every implement dealer in America offered these for sale in the parts department.

Auxiliary warning lamps were available on CAs. The same lamp was used on WD-45 tractors. The new attachment was authorized May 28, 1954. The assemblies were made by Guide Lamp.

Both Delco and Auto-Lite generators were used on Bs, Cs, and CAs.

A Masonite pressboard insulator was located between the top of the battery and cover. On the IB, a spruce insulator went between the battery and the battery box side.

Steering Gear

The steering mechanisms in the B and the C are not interchangeable. The B worm shaft has a left-hand cut, while the C has a right-hand cut. Steering gear parts are now difficult to get, especially for the Gemmer brand. Restorers looking at projects should take steering gear condition into account when considering a project. Move the steering wheel back and forth to see how far you have to move the wheel before the front wheels start moving. Any more than a quarter turn indicates excessive wear.

CAs used steering wheel 225330, covered by a large acorn nut. The wheel was made by Sheller and was 15 7/8 inches in diameter. It was originally painted orange, but the drawing was changed on January 17,

Most Cs came with electric starting, but you could still hand-crank them. The B/C engines can be a little finicky about waking up, and sometimes you do have to get out the crank. The air intake cap would, of course, be orange on a freshly produced tractor.

This B has an original-looking oil filter but a replacement coil, which are often replaced and left black. The distributor has been painted orange, with the data plate properly masked over—they wouldn't have put them on if you couldn't read them after being painted at the factory! The starter is painted black.

This otherwise unusual dual-transmission Model C has a fairly original looking electrical box located above the top of the extra gearbox. Many electrical boxes have been modified by cutting them and putting in a more modern (and cheaper) toggle switch.

1956, to all black. It had a small Sheller medallion on the wheel, on a spoke near the rim.

Two different makes of steering gear were used on the B and C family tractors. Gemmer gear was used on B-101-46120, except for B-29121-30611, B-31045-35327, and B-38913-42251. The Gemmer gear was also used on all C tractors before C-42611. Ross Gear was used on B tractors before 101, the exception tractors noted above, and on B-46120 and after, and C-42611 and after. Both steering gears were apparently also used on IB tractors.

The early Bs had a steering knob as original equipment but that was discontinued, probably before 1940. Bs before Serial Number 101 used Ross steering gear and steering wheel 205623, which was replaced by 207562. At least some, if not all, of the steering wheels were black before 1940, but a note was added to the parts drawing in December 1940, specifying that the vendor paint the wheel orange. There is some reason to believe that this steering wheel may have changed from orange to black at the same time that the CA's wheel did in January 1956.

Wheel and Tire

The pneumatic rear wheels were dished on some B, C, IB, and RC tractors. Several different treads could be obtained by either reversing the wheels or moving the rim lugs. There were several options available to go beyond the normal range of treads as well, either narrower or wider.

On the CA, Allis-patented adjustable wheels were used. Known as "Power Shift" wheels, these were a godsend to farmers with small vegetable crops, with multiple row spacings. Lugs rode on rails welded to the rims. To move the row spacings, the farmer unbolted the lugs and put the tractor into gear. As the rear wheel moved, the lugs forced the sloped rails in or out, moving the whole rear tire.

Several front axles were available with the B. Standard was a nonadjustable wide front axle. The

CAs came with Power Shift rear wheels (with silver rims), a good idea for any cultivating tractor. CAs also came with hydraulics with Traction Booster.

front wheels were dished, so that small adjustments could be made by reversing the wheels on the hub.

The CA had single front wheel, dual-wheel narrow front (standard), and a wide-tread adjustable front axle. Wheels were the same on the adjustable wide front and dual-wheel narrow front, but a cast wheel was used on the single front. The single front wheel used a 6x12 tire, while the other two options used a 5.00x15 tire as standard.

Model C Specifications from the Advertising Literature

Overall Dimensions—Wheelbase, 77 inches; length, 114 inches; width, (narrowest tread) 68 1/2 inches, (width widest) tread 89 1/2 inches; height, 62 3/4 inches; ground clearance, 23 1/4 inches; turning radius, 7 feet.

Regular Equipment—Rubber tires, starter, lights, muffler, fenders, thermostat, radiator shutter, cushioned seat.

Special Equipment—Oversize (10x24) tires, front and rear wheel weights, wide adjustable front axle (factory only), swinging drawbar, side hill hitch, hand or hydraulic lift for tractor-mounted implements.

Front Axles

The Bs had several different front axles available. The early front axles on the tractors before B-101 were apparently hand-forged to a slightly different curve than on the later tractors. Spindle arms, link assemblies, and other parts were different as well. Standard Bs 101 and after used a nonadjustable arched axle. Asparagus tractors used a nonadjustable front axle with a different center arch and spindles.

Two different types of adjustable-tread front axles were available with Bs. The first type, used prior to 1940, had a shallow, squared arch to the center section of the axle. The second type, used in 1940 and after, used a straight, square steel tube section, with square spindle supports ("axle extensions"). Axles built in 1940 and 1941 were adjustable from a 38-inch tread to 60 inches. Axles built in 1942 and after were adjustable from 40 inches to 65 inches.

IBs used a straight nonarched nonadjustable front axle. Parts were different from the B except for spindles. There are two different pivot pins listed for the front axle, one for use with a main frame (certainly the IB) and one for use without a main frame, probably for the rare Allis B Orchard tractors.

The C standard front axle was the two-wheel narrow front, which could be adjusted by reversing the wheels on the hubs. In addition, a single front wheel was available with a forked wheel support.

An adjustable front axle was also available. The adjustable front axle option was adjusted in the same manner as the Bs. Two retaining bolts could be removed, clamps on draglinks loosened, and the spindle support tube could be moved in and out of the main axle tube. The wheels of these axles were also dished, and could be reversed on the hubs to provide extra treads as well.

CAs were available with the narrow tricycle front, single-wheel front, or the adjustable wide front axle shown here. This tractor has the distributor ignition that was used on engine CE-149840 and after, that is a very popular retrofit to older, magneto tractors. Note the unpainted data tag on the distributor.

Bs with nonadjustable front axles used one-piece curved forgings. There were several different versions of this axle, including the preproduction (before Serial Number 101) axles that were hand-forged, and versions for narrower asparagus tractors. This is the "normal" axle.

This is another view of the Model B axle, from the front underside of the tractor. The whole purpose of the B can be seen here. A single row passed down the center of the tractor, under the high part of the arch, then between the two dropped final gear housings on either side of the rear axle. The drawbar would have been removed for cultivating, and cultivator gangs would have been located on the right and left sides of the tractor, front and back.

Allis Cs had as special equipment a "Flip-Over" seat that included a shock absorber. A shock-absorbed seat assembly was used on the CA. A sheet-metal pan seat was used.

Brakes

Hand brakes were used on Bs before serial 52718, C-5756, and IB-1001. One lever was used on each side of the operator's platform to assist in turning the tractor, which would have made road travel exciting! On tractors before B-1101, a separate knob at the top of the brake levers was used. On later Bs and Cs, a knob integral to the lever was used. A hand lock was used at the base of the levers to lock the brakes. A machined yoke with a pivoting lever was used on tractors before serial B-101, while on later tractors the yoke was integral to the lever, which was a piece of stamped sheet metal.

IB tractors before IB-1001 used two hand brake levers that were made of bent flat steel stock. The IB brake system had a central foot pedal to engage the parking brakes.

On B-52718, C-5756, and IB-1001 and after, foot brakes were used.

The small size of the tractor, and the desire to keep it light and cheap, led Allis to combine the belt pulley and the PTO into one unit on the Model B. The curved drawbar moved the hitch point out in back of the tractor, while providing multiple hitching options.

PTOs and Belt Pulleys

Three different PTO arrangements were used on B, C, and IB tractors. The first PTO assembly did not include a hydraulic pump and was used on B and IB tractors only. The second version, used on all three tractors, had a hydraulic pump drive and a 1 1/8-inch nonstandard PTO shaft. The third style PTO had hydraulics, but used the ASAE Standard 1 3/8-inch PTO shaft. PTOs without hydraulic pumps used a different engagement lever than the hydraulic tractors. A PTO warning decal was located on the sheet-metal guard assembly.

PTO guards on Bs and Cs were open cylinders before 1940. Then, the guard formed of sheet metal with braces was used until the auxiliary warning light sockets were added in the 1950s. The later guard, 226867, differed only in having two extra holes for mounting the auxiliary warning light.

The PTO housing also supported the belt pulley assembly. Belt pulleys used before 1939 were cast iron. Belt pulleys used in 1939 and after were steel. Most CAs used an aluminum belt pulley weighing about 5 pounds, which reduced gear noise.

The IB's PTO housing on the left has the PTO warning decal in about the right spot. The drop housings are rotated 90 degrees forward to lower the tractor. The IB hydraulic control linkage can clearly be seen (the operator's lever is out of view on the left side of the photo).

Drawbars

Two drawbars were available for Bs. The regular U-bar drawbar bolted to the drop final drive housings and had six adjustable heights. A clamp went on the outside of the drawbar where it bolted to the housings.

The ASAE drawbar was designed to fit implements manufactured with ASAE standard distances between hitches and PTOs, which guaranteed interchangeability between manufacturers. Two steel braces dropped from the corners of the fender braces

This is a fairly standard rear end of a C. Notice the green seat, the U-bar drawbar, the PTO warning decal, and the warning lamp. The PTO/belt pulley unit is tucked in under the seat on this tractor.

Model CA Specifications from the Advertising Literature (Form TL-5511)

Dimensions—Wheelbase, 81 7/8 inches; overall length, 124 5/8 inches; height at muffler, 76 3/8 inches; narrowest width, 67 3/8 inches; widest width, 90 3/4 inches; turning radius, 7 feet.

Standard Equipment—Starter and lights, hydraulic system with Traction Booster, Snap Coupler hitch, Power Shift wheels, Two-Clutch Power Control, ASAE drawbar, belt pulley, PTO.

Extra Equipment—Wheel weights, front-end weights, remote hydraulic ram.

to the drawbar to support it. A triangular plate was added to the U-bar drawbar to extend the hitch point a few inches to the rear to match the ASAE standard. There was only one height to use with this drawbar, due to the braces.

Hydraulics

The hydraulic system on the B, C, and CA tractors was set up for 3,200 pounds of pressure at the pump, much higher than other tractor manufacturers'

Some CAs were fitted with Allis' unique Snap Coupler hitch. This hitch was much easier to hitch than a three-point—the operator backed into the implement to "snap" the lower arm, then attached the two top arms seen here. The Snap Coupler lost out to the three-point when other manufacturers followed Harry Ferguson's lead.

systems; the principle was high pressure, low volume. Allis' rams were much smaller in diameter than those of competing firms. When using today's larger hydraulic rams, make sure hoses are in good shape; otherwise, they'll let loose from the pressure. Large cylinders can result in too much fluid leaving the tractor and going into the ram, producing a loss of lubrication. Gauges must also be able to handle the pressure, which has made it difficult to find new reproduction gauges. The fluid drain plug on the bottom of the PTO unit on the CA was magnetic to capture metal particles in the system.

There were at least three different hydraulic control handles on CAs. Lever 225651 was used on tractors CA-1185 to 25124 (before pump 1182), while 225920 was used on tractor 25124 and after. Both pumps used rubber knob 220901 on the arm.

Snap Couplers

Some later Bs, and at least some CAs (Serial Numbers 31455, 31319 and 20, and 3142331444, and possibly more) used a Snap Coupler hitch assembly. Snap Coupler was introduced in January 1954. The system used a bell-shaped housing under the tractor with an automatic trip. The operator backed up to an implement, which had a point that slid into the bell housing and automatically latched. On the lift arms for the hydraulics, a latch was used that had to be triggered by hand. To unlatch the bottom hitch, a lever was used, located at the operator's feet.

Weights

Front weights were available on the B, IB, and C tractors. The weights attached to either the front bolster or the adjustable front axle. The weights looked like an "M" shape.

Rear wheel weights for CAs came in two varieties, inner wheel weights that attached directly to the wheel, and outer wheel weights that attached to the inner wheel weights. Multiple outer wheel weights could be used as long as they attached to the inner wheel weight.

Chassis weights (identified as "engine side weights") were also available on the CA.

The Allis Snow Tractor

Allis-Chalmers started releasing specifications for the "Snow Tractors" in September 1942, and continued experiments into 1943. The tractors were identified as both T-26 and T-27. A variety of different combinations of tracks and engines were tried, including long- and short-shoetracks, Harley Davidson engines, and Hercules ICK engines with both aluminum and cast-iron blocks. Allis also manufactured the "One Ton Snow Trailer." The main production variant was the T-26 with the Jeep engine.

Unlike the WD and WD-45 tractors, the CA did not have the hydraulic pump instruction on the right-hand side of the battery box, but on the inside of the left fender (hidden in this view). The inclined rail construction of the Power Shift wheels can be seen here. To adjust the rear wheel width, the operator loosened the nut and drove backward or forward. The motion of the wheel against the curved bars of the rim forced the wheels in or out.

This Snap Coupler latch assembly was used on tractors built after October 1955. The Model CA in these photos was built in 1953, but the two-part stamped Snap Coupler assemblies bent up easily, and many have been replaced by the more modern, heavier-duty part. The latch assembly is the part above the right end of the spring in the left-hand part of the photo.

The RC was intended probably as an attempt to compete with the Farmall F-12/F-14 small two-row cultivating tractors. The tractor had a WC frame with a slightly enlarged B engine (known as the "CR" engine). The result was less than successful in sales.The RC's main identifying feature is the small engine and hood on top of the large frame, which leaves a lot of space between the front pedestal and the radiator. Some parts are unique to the RC, basically those parts connecting the frame to the engine, or controls dealing with the different geometry of the tractor. The RC is hitched to one of the last Roto-Balers ever produced, built in the 1970s but never assembled until the current owner purchased it from a dealer.

This view clearly shows the B heritage of the engine. The air cleaner has the Donaldson air cleaner decal used in this vintage of tractor with the round "Donaldson" air cleaner logo between the "Service Daily" wording. The muffler is the "squashed tube" that was very common for Allis at that time.

The proud owner had highlighted the serial number of his tractor in white. Allis serial numbers seem to have been ground out for several occasions, including dealerships shuffling tractors among themselves in violation of Allis-Chalmers corporate policy and other less-legal reasons. Some dealers eradicated serial numbers of early tractors that were traded in to make them more attractive to customers who knew that early tractors hadn't necessarily had the bugs worked out of them.

An original oil gauge is seen here, along with the steering shaft support. The steering universal joint would probably have gone just under the support.

The RC's combination of a large, expensive frame with a small, less powerful motor created an expensive small tractor, although its ability to use WC implements like cultivators must have saved Allis some tooling money. The RC remained in production for only four years (1938 saw three built before larger production began in 1939). This is Serial Number RC-9.

RC Tractors

The RC was essentially a mixture of WC and B parts put together in an attempt to make a cheap two-row tractor to compete with the IH model F-12 and the John Deere model B. What Allis got was an expensive and underpowered tractor. The RCs are fairly rare, and a nice tractor for a collector—a little lighter than a WC, yet a more imposing tractor than a B or C, and with a certain "cute" factor.

Engines

RCs used an R engine that was essentially a larger bore B engine. Eventually, Allis designed the "CE" engine that went into the B, C, and CA tractors which used the larger bore.

Chassis and Transmission

Most of the rest of the tractor was made up of WC parts, with different engine mounts, clutch linkages, etc., to link the R engine with the WC frame.

Steering Wheels

There were several sources for the steering wheels for the RC. Diameters included 15 7/8, 16, or 16 1/2 inches. The wheel was painted orange after 1940, with some earlier steering wheels possibly being black.

Front Wheels

Front wheels generally used 15-inch tires, although some other sizes were possible.

The decal seen on the RC's frame rail is the steering universal joint lubrication instruction. It probably should be on the shaft itself, just below the support, as it was on the WC. Unfortunately, the information describing the decal locations for the RC wasn't found. The oil filter is a more modern Allis replacement. The original filter would have been the Allis can type. For this era of tractor, the front wheels may have had a stripe on the rim, but the wheel/rim drawings for the RC were not examined.

This is where the RC's starting crank got stowed, on the inside of the left fender with clips, although in actual use, many of them ended up in the toolbox. The clips did eventually stretch out, failing to hold the crank as well as they could, resulting in lost cranks. The handbrake lever is also seen here.

The Allis part of the snow tractor can best be seen in the front grill. Other parts included an outside engine. Jeep, Hercules, and Harley-Davidson engines were all experimented with, but the Jeep engine won out for production. The rubber tracks were also used by Studebaker Weasel tracked vehicles made for the military. The rubber tracks, now over 55 years old, have deteriorated with age and are unreplaceable, meaning this snow tractor stays on the trailer

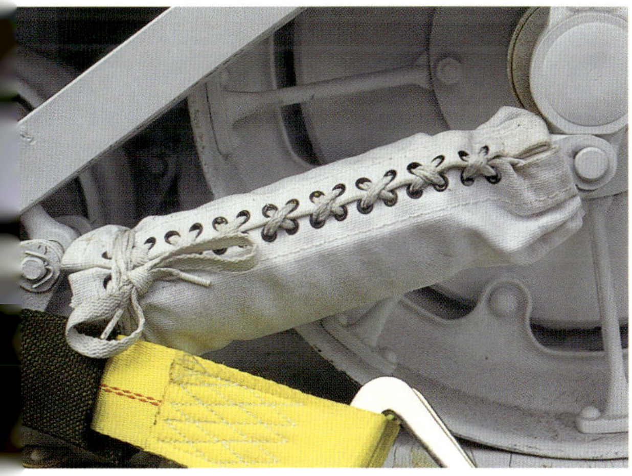

Getting the machine to function in a very hostile environment was challenging, and required attention to many small but crucial details. This boot is one of them—it keeps snow out of a suspension spring. Without the boot, the snow gets into the spring, builds up, and is compacted into ice, freezing the spring up and giving a rough ride.

Allis-Chalmers snow tractors were designed to be used by the U.S. Army for missions in cold, snowy climates. Reportedly, the main goal was for Search and Rescue in Alaska for pilots ferrying aircraft to Russia under Lend-Lease during World War II.

The front view of a G shows the primary goal of the tractor's design: a small tractor with an unobstructed view for the operator. The working implements of the G mount between the operator and the front wheels, under the frame arch. The only better way to see what you are doing is to plant and cultivate with a hoe.

CHAPTER 8

Model G

Although Allis-Chalmers had just a few years earlier pioneered a whole class of small tractors, by the end of World War II, Allis again took tractor design smaller, producing the radically different G. Allis didn't produce several imitations the way it had with the B. While several other very small tractors would be produced in the same years, they would be designed with conventional formats. The G, however, was imitated, and the basic layout of the tractor is still being used today.

Produced at Allis' new Gadsden, Alabama, factory, the G was unlike anything else on the market at that time. In fact, the tractor harkened back to an earlier age. Closely resembling the IH Motor Cultivator, the tractor featured an engine, transmission, and drive wheel unit in the rear of the tractor, and a space frame forward, giving an open view of the cultivator directly in front of the operator. A variety of implements were specially made to fit the unique frame of the G. Allis purchased many of the parts outside the company, including the Continental 62-cubic-inch engine.

For a brief time in 1982, Allis sold a short run of "GII" tractors in Canada under the Simplicity name. These tractors were apparently produced by an outside supplier. Several companies over the years built Allis G copies, including some in Wisconsin.

Engine

The engine was a Continental Y-62, similar to some Massey-Harris engines. Many collector believe that the engine is also similar to a Farmall Cub engine, but the Cub engine was built by International Harvester.

Carburetor

A Marvel Schebler TSV-13 carburetor was used.

Air Cleaners

An optional air cleaner stack assembly was available for dusty conditions. The assembly consisted of a stack extension, a cap, and two braces. The air intake was raised to get it out of the "dust zone" of the tractor and into clean air.

Cooling System

Gs used a thermosyphon cooling system that had no thermostat, radiator shutter, or temperature gauge. Keeping the radiator full of coolant was critical. If the coolant level fell below the intake, the natural flow was lost and the engine overheated. The cooling system was pressurized and used a pressurized cap.

Transmission

Gs were available with a special low gear that gave a speed of 0.80 miles per hour at 900 rpm. This was another gear in the transmission, rather than a different gear ratio. On tractors without the special gear, the gear position was filled by a spacer.

The G's gear shift assembly was the same as on the B and C tractors, except for the guide plate, notched for the special low gear. In addition, the lever was shaped differently and guide pins were used.

Rear Axle

The rear axle housing assembly was a casting welded to a shaped tube. In farming practice, Gs were often narrowed in the field by taking a section out of the rear axle housing tube, and taking a similar section

Gs are popular with a lot of people. For some reason, many G collectors collect of lot of them, despite their rarity. The basic layout of the arched support leading to the torque tube, followed by the rear axle in front of the engine, created a well-balanced, small, light tractor that was nearly perfect for small farms and specialty crops.

The rear view of the G again shows the amazing compactness of this tractor. The narrow body was designed to go down between rows. Although not normally thought of as a drawbar tractor, the G was fitted with a drawbar, and with the weight centered near the rear axle, it has quite a bit of pull. The author witnessed a G pull at the Hutchinson, Minnesota, show using a garden tractor sled. The Gs pulled the sled with great ease, despite their low horsepower.

The Allis G was a novel idea from a company that didn't have many restrictions on thinking in the 1930s or 1940s. This basic design of the tractor, although seemingly so unusual and impractical, is still copied today.

This tractor has a metal dealer's plate, where most dealers used a decal. Preserving the individual touches a tractor received can be difficult but rewarding at shows where the extras draw attention and create conversation. Data plates on the generator and starter are left unpainted. Most, if not all, archival photos show G generators and starters to be orange like these. The hose was also painted orange.

The Gs used an unusual flat muffler to keep their profile low. The taillight on this tractor has a wide chrome bezel. Other G taillights have narrow bezels, but it is unknown when the change was made.

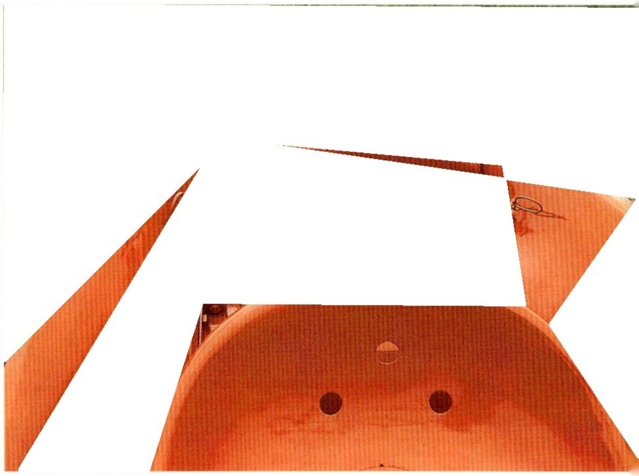

ABOVE
A view toward the rear of the tractor over the operator's position shows the loops for the starter, on the left, and the choke, on the right. The front cap just behind the seat is the fuel tank cap.

LEFT
Another view of a different Allis G shows a different taillight assembly, this one without a chrome bezel. This G also has the special air intake extension that was sold for dusty conditions. The rod for the choke can be seen running from the fender to the carburetor. The curved pipe is the air intake running from the air cleaner on the right side of the tractor.

This shows what is known as the "new style" belt pulley attachment (the pulley itself has been replaced by a V-belt pulley). Just forward, the left axle assembly shows the construction of a tube welded to a casting. The oil pressure gauge is next to the muffler.

out of the rear axle, and changing the front end as well. Entire fleets of Gs had this treatment, especially on large vegetable farms that operated many Gs at once. One Allis-Chalmers magazine described an operation in California that had 13 Gs modified into narrow-row tractors by the owner.

Belt Pulleys

Two styles of belt pulley assemblies were used. The old style had a drive running through the assembly for the hydraulic pump. The second style, which was first produced in 1950, did not have the extra drive, but was a straight shaft running in a housing.

Front Wheel

A 4.00x12 tire was standard.

Rear Wheel

A 6x30 rear tire was standard. Rim assembly 800545 was used.

Drawbars

Two different drawbars were used on Gs. The early drawbar had a tang that slid on top of the support,

This photograph shows the "old style" combination hydraulic pump and belt pulley assembly (again, the pulley itself has been replaced by a V-belt drive pulley). That's an original oil gauge, marked "Allis-Chalmers," on the right near the choke rod.

A look forward from the operator's seat shows the excellent vision available for cultivating, planting, and other jobs the G could perform. The unusual steering wheel was unique to Gs, and was furnished from the factory looking like this.

but did not fully go around the support. The later support, used on Serial Number 26751 and after, had the tang that went all the way around the support.

Steering

The G used a uniquely shaped steering wheel, flat on top and round at the bottom. The wheel was painted Persian Orange.

The brutal simplicity of the G is again shown in this view of the front end. The simplicity probably contributed to the large numbers of companies copying this unconventional tractor.

Electrical System

Gs came fitted with a complete electrical system. A small panel held the ammeter, which was the standard Allis part number 208302. The generator cut-out relay also attached to the panel bracket. Tractors before 20972 had a wiring harness clip that attached to the generator bracket. On later tractors, the wiring harness was routed behind the oil filter. The system was 6-volts. The starter pull rod was located behind the operator on the right side, and had a small loop in the end. A wooden battery insulator was used in the battery box.

A single headlight was used on the G. A single small taillight was used, with a ruby glass lens. The wiring to the headlights on the front frame of the tractor ran through the torque tube. The distributor was a Delco-Remy model 1111708.

Hydraulic System

A G hydraulic pump puts out about 900 psi, and was mounted on the side of the belt pulley assembly. The hydraulic control rod had a knob at the end, 220901.

Hydraulic Lifts

Two different hydraulic lift shafts were used. The solid type lifted one set of implements while the split type lifted two implements, having a rod running through the main rocker arm tube attached to another arm assembly that acted independently. Two trip levers were used to disable one arm at a time.

Front Weights

A front weight was available for the G that was made of cast iron and was triangle-shaped to bolt over the extreme front nose of the tractor. There are modern reproductions of this weight being sold, but these are cast in aluminum.

The hydraulic pump on the G was intended for low flows, so a PTO-driven hydraulic pump was added. These pumps are a common add-on to many tractors that have had aftermarket tools or hydraulic systems added.

This is apparently an aftermarket hydraulic valve and reservoir. The small-capacity, high-pressure system of the original Allis hydraulic system would be hard pressed to handle the loader. The control handle for the original hydraulic system for the Allis G was a long arm with a knob at the end, similar to the CA.

Adding hydraulics rapidly increases the complexity of the G. This is the "solid type" of hydraulic lift, showing the narrow diameter of Allis hydraulic rams of the era. Parts of the cultivator attachment are mounted on the lift.

The normally low-profile G gets a lot taller with a bucket. The G can't really stand up to too much of a load in the bucket, and really shouldn't—a tip-over could easily happen with a light, short tractor under a tall, heavy load.

Due to the light weight of the G front end, Allis offered a weight attachment that fit over the steering box. The weights are being reproduced in aluminum, while the originals were made of cast iron. The weights are very useful for those brave souls pulling with Gs.

LEFT
The front end of an Allis G is probably the simplest of any tractor, period. The light should probably be painted black, the way it came from Guide Lamp. The wire runs into the torque tube.

CHAPTER 9

Wheels, Rims, Paint

It is appropriate to list together wheels, rims, and paint, due to the way that Allis, like other companies, arranged its paint schemes and painting operations in the factory. Starting roughly in 1930, Allis' new tractor assembly line, built primarily to produce the U and UC tractor (but also the 25-40), completed and in some cases tested the tractor before painting it orange. When painting the tractor, some parts were either masked off or removed from the tractor to avoid the paint. While the listings for what did not get painted orange are now missing, the following items have been identified, either from photos or from the part drawing and material specification itself. Other items are identified by inference from later painting instructions from tractors built in the 1960s.

Painting Instructions

The following parts were not painted orange:

- Wheels and rims and pneumatic tires
- Steering wheels
- Rear taillight casings (left black)
- Headlight casings
- Plated hardware
- Glass parts—gauge faces (and bezels), headlight lenses, sediment bulbs, glass oil filters (bezel colors are noted in the parts section)
- Seat upholstery
- Spark plugs
- Generator/starter

Examination of photos shows that the following parts were painted orange:

- Spark plug wires
- Magnetos
- Shift and brake lever knobs
- Manifolds (although the data plates were not)
- Pipe-type mufflers (although the paint burned off quickly)
- Wiring harnesses where exposed
- Unplated hardware

There is some variation to the items masked off or left unpainted. Photographs taken in the 1950s show some, if not all, CAs with orange-painted generators and starters. A few WD-45 photos have been seen with this feature. Steering wheel paint specifications also changed from orange to black on CAs, the WD-45, and the Bs in 1956.

The paint was Persian Orange, now known as "Persian Orange Number 1." The orange color that Allis used subsequently was changed periodically over time, and today's AGCO Allis tractors are a quite different shade. The Persian Orange Number 1 seems to be very sensitive to various changes in paint chemistry and applications, so it is common at shows to see several different shades of orange in a line-up.

Wheels and Rims

A bewildering variety or wheels, rims, tires, and lugs were available from Allis-Chalmers during this period. The author originally tried to come up with a way to accurately and simply identify the equipment and varieties. However, the various things that Allis did, combined with some problems with the records, made it impossible to do the job without instilling a false sense of confidence in the reader about the completeness of the job. A true accounting of the equipment based on existing records would be voluminous, and it would be incomplete.

Orange Allis tractors with steel wheels seem to always have orange steel wheels, which were probably painted along with the rest of the tractor.

Model As seem to have black front and rear wheels, although given Allis' changes and variety of suppliers, other variations may be possible.

Allis generally made available all possible combinations of tire equipment, i.e., all-steel wheels, steel fronts with air rears, air fronts with steel rears, and all air. In industrials, solid rubber equipment was generally available. In addition, Allis shipped many tractors with no tires or wheels at all. These tractors were usually bound for three markets—export, where local equipment was available (saving shipping, money, and sometimes meeting local import regulations), manufacturers using Allis tractors as bases for their own equipment (including Thompson), and domestic U.S. ag users who needed or preferred local or special wheel equipment, including farmers in Florida and California. So in reality, even brand-new tractors could have almost any tire equipment under the sun.

Wheels and rim colors varied considerably. Allis-Chalmers bought its pneumatic tires and rims, often in

WCs were available with a large variety of different wheel and rim colors for the air tire tractors, including, on the unstyled tractors, green. Orange rims front and rear were just one of the combinations available on styled WCs.

WDs are usually seen with all-orange fronts and orange wheels and silver rims on the rears for their tired tractors. However, some WDs with non-Power Shift rear wheels may have had orange rims.

large batches. A change made in color, although it may have been quickly overturned in the production records, may have resulted in a large order of rims or wheels produced in a certain color that covered production for quite some time. Often, rims and wheels had multiple sources of supply, with the manufacturers having different processes and coating techniques. Also, Allis gave the manufacturers some leeway in determining how a particular rim was to be supplied. In short, there could have been three tractors coming down the line in some years in the 1930s, identical except for three different rim colors! Some wheels and rims did not have color specifications, presumably being painted by Allis itself to unknown specifications.

Parts that were originally produced for tractors built before 1958 but continued on D-series or beyond tractors often had their paint specifications changed to match the new paint schemes. If you were to go to a dealer and buy a front wheel hubcap for your 1954 tractor, you would almost certainly end up with a cream colored hub, due to the part being used in D-series production and part of an order for producing those tractors.

Steel wheel equipment does not seem to have special painting instructions listed on the parts drawings, indicating that the wheels were painted Persian Orange with the rest of the tractor. It would not surprise the author to learn of exceptions to this idea.

Nuts and bolts on wheels and rims were probably hand-painted to match on earlier tractors. However, in November 1950, it was noted that CA tractors were now receiving cadmium-plated hardware to avoid hand-painting.

It should be noted that sales literature and parts books do contain information about wheels and rims. Moreover, the equipment listed in the manuals and literature was usually only a portion of what was available, or what certain special customers could obtain.

Model A Wheels and Tires

Originally, the front rims were black. In March 1938, the option was given to the manufacturer to go to black or galvanized. Goodyear-built rims could be provided to Allis in Bond-A-Coat. I did not find the rear rim drawings, but they almost certainly followed suit.

Model U and UC Rear Wheels and Tires

Rim 204178 was black until July 1936. Color was then changed to a "natural steel finish, painted black on inside of the rim only. Color was changed to all black in March 1938. Eventually, rims purchased from Goodyear were either enameled black or galvanized, with Bond-A-Coat optional. Rims purchased from Firestone had aluminum paint with Persian Orange wheels, although a few wheel and rim assemblies may have continued to be black through 1952. IUs had an extensive range of available wheel equipment—at one point, Allis released 16 different new wheel, tire, and rim options in a single day.

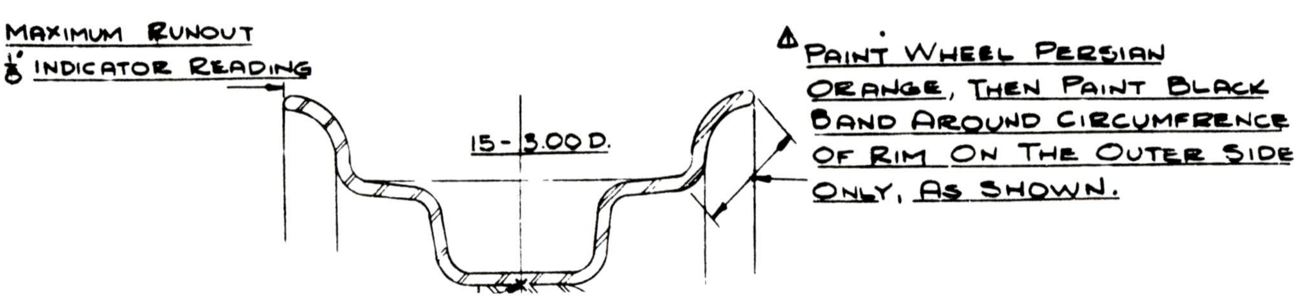

Bs had at least four different paint combinations for front wheels in the 1930s, including orange wheel with all-black rim, orange wheel and range rim with wide black band (seen here), orange wheel with orange rim with two narrow black bands, and orange wheel with orange rim with one narrow black stripe. Eventually, all-orange rims, and possibly all-silver rims, were used as well.

Model U and UC Front Wheels

U fronts generally changed to match the rears. One unusual tire for UCs was a solid rubber front tire that was apparently an option for a short time.

Model WC, WD, WF, and WD-45 Rear Wheels

WCs of 1933 probably had Persian Orange rear wheels. In January 1934, the paint color was changed to olive green, which lasted on wheel orders to April 1934. In April, the color was changed to black.

The WD and WD-45 with the slide-out rears had orange wheels and aluminum-dipped rim and rail assemblies.

Model WC, WD, WF, and WD-45 Front Wheels

The original color was Persian Orange, changed in late February 1934 to olive green. This was changed in April 1934 to black. Black seems to have lasted quite some time.

In the WD era, at least two different combinations are noted in the parts drawings. Some wheel and rim combinations are all orange, while some have an orange wheel and aluminum rim. The color may depend on which equipment you have. It appears that single fronts were usually orange.

Cs had some of the variability of the Bs, including orange and possibly silver front rims, and probably had orange and silver rear rims also.

Model B, C, and CA Front Wheels

The original B pneumatic front wheels had an orange wheel with black rim. On some, if not all, wheels, the color was changed to all orange with a black stripe on the rim (208078). A majority of the tractors with Serial Numbers 9019–26525 used a front wheel that was Persian Orange, with two black stripes, as shown in the drawing. Stripes were removed at some time in the 1942–1945 era, but the date was not kept. Front wheels and rims were usually orange, although others may have had aluminum rims in later years.

Model B, C, and CA Rear Wheels

On one type of wheel rim authorized in 1941 (4.00x30 crop center rims for the B Vegetable tractor), the note said that the manufacturer was to enamel black or galvanize.

On CA tractors, the slide-out rims were dipped and sprayed in aluminum paint, with orange wheels. There were two different manufacturers, Firestone and Motorwheel. The rims they manufactured were slightly different, but Allis specified that the rims put on a particular tractor must come from one source—mixing and matching was not allowed. The note may have been added, however, after some mixing and matching had occurred.

The author didn't get a chance to look at G rims closely, but silver seems to be the most common color, with some orange rims showing up.

APPENDIX

Serial Number Information

Allis serial numbers generally started with Serial Number 1, though there were exceptions. Some tractors had several preproduction versions that started with 1, with production tractors starting after those numbers.

Tractors with consecutive numbers were not necessarily built one right after the other. If a tractor was built to order and parts were specified that were not in stock, the serial number was held until the parts came in and the tractor could be built. A tractor was not built for every serial number in the case of cancellations or other problems.

Tractors shipped for export out of the United States had to be marked "Made in U.S.A." There were two ways of doing this. Patent decals could be used (patent decal 213538, used originally in 1942, had this information). If the equipment to be exported didn't have this decal, the words "Made in U.S.A." were stamped on the machinery near the serial number.

WD-45 and WD-45 Diesels used the same range of serial numbers. "D" was added to the serial number on diesel tractors.

Serial numbers get obliterated on tractors for several reasons. Dealers sometimes obscured serial numbers when they reallocated scarce tractors among themselves after the war. When the company tried to catch the dealers, it couldn't prove anything without the serial number. Farmers obliterated the serial numbers as they tried various tax dodges and serial numbers were erased by thieves. Since Allis stamped the numbers directly into the cast iron of axles and transmission casings, numbers can sometimes be read, due to the fact that the stamping process caused certain stresses in the surrounding numbers. State crime labs, police departments, and possibly certain private detectives and insurance firms can perform tests to read these numbers. Contact your local police department to see who in the vicinity can perform these tests.

B and IB tractor serial numbers are stamped on the transmission case next to shift pattern. The engine number is stamped on the flange where the engine mates to the torque tube on the left side of the tractor, between the air filter and the sediment bulb.

WD tractor serial numbers are stamped either at the left brake cover or the left rear of the axle housing. The engine serial number is stamped on a pad on the left side of the engine just under the hose running from the air cleaner to the carburetor.

G serial numbers are stamped on the transmission case just in front of the gearshift. The engine number is stamped just in front of the starter.

Tractor Serial Numbers

Model E 25-40

Year	Serial Numbers
1930	24186–24842
1931	24843–24971
1932	24972–25023
1933	25024–25061
1934	25062–25308
1935	25309–25581
1936	25582–25611

(E engines and skid units produced into the 1950s for Cooper; power units produced for other customers.)

Model A

Year	Serial Numbers
1936	25701–25725
1937	25726–26304
1938	26305–26613
1939	26614–26781
1940	26782–26895
1941	26896–26914
1942	26915–26925

Model U and United

(Includes UI, U orchard, UE skid units for Cooper, other tractors.)

Model Us with Continental Engine

Year	Serial Numbers
1929	1–1974
1930	1975–6553
1931	6554–7261
1932	7262–7404

Model Us with UM Allis engine

Year	Serial Numbers
1932	7405–7418
1933	7419–7684
1934	7685–8062
1935	8063–9470
1936	9471–9988

Model Us with UM Engine, 4 1/2 inch bore

Year	Serial Numbers
1936	12001–12821
1937	12822–14854
1938	14855–15586
1939	15587–16077
1940	16078–16721
1941	16722–17136
1942	17137–17469
1943	17470–17801
1944	17802–17819
1945–1946	no information
1947	18309–18612
1947	20174–20178
1948	18613–19421
1948	20179–20300

1948	21000–21305
1949	20301–20421
1949	21036–21986
1949	22018–22023
1950	20422–20512
1950	22024–22547
1951–1952	no information

(U components were produced for some years afterward for other manufacturers.)

Model UC

1930	1–38
1931	39–1099
1932	1100–1231
1933	1232–1293
1934	1294–1551
1935	1552–2000
1936	2001–2770
1937	2771–3756
1938	3757–4546
1939	4547–4769
1940	4770–4971
1941	4972–5037

UC Cane

1944	5038–5067
1945–46	None
1947	5068–5267
1948	5268–5525
1949	5526–5643
1950	5644–5805
1951	5806–5938
1952	5939–6142
1953	6143–6217

(UC skid units produced for many years afterward (at least to 1960) for use in the Thompson tractors, especially Model X tractors.)

Model WC

1933	1–28
1934	29–3126
1935	3127–13869
1936	13870–31783
1937	31784–60789
1938	60790–75215
1939	75216–91533
1940	91534–103516
1941	103517–114533
1942	114534–123170
1943	123171–127641
1944	127642–123623
1945	123624–148090
1946	148091–152844
1947	152845–170173
1948	170174–178202

(WC skid units were produced for several years afterward for other manufacturers.)

Model WF

1937	4–288
1938	289–1335
1939	1336–1891
1940	1892–2299
1941	2300–2703
1942	2704–3003
1943	None
1944	3004–3194
1945	3195–3509
1946	3510–3747
1947	3748–4110
1948	4111–5499
1949	5500–7317
1950	7318–8315
1951	8316–8353

Model WD

1948	7–9249

(First seven are probably preproduction tractors. First production tractors are identified as "WD–2" tractors, indicating possibly a small run of other tractors.)

1949	9250–35444
1950	35445–72327
1951	72328–105181
1952	105182–131242
1953	131243–146606

(WD skid units produced for at least a short time, after the end of regular production, for other manufacturers.)

Model WD-45

1953	146607–160385
1954	160386–190992
1955	190993–217991
1956	217992–230294
1957	230295–236958

Model B (Includes Early IB to 1945, B Orchard)

1937	1–96
1938	97–11799

(However, parts books usually list the regular production Bs starting at 101, so there may be a small gap between 97 and 101.)

1939	11800–33501
1940	33502–49720
1941	49721–56781
1942	56782–61400
1943	64501–65501
1944	65502–70209
1945	70210–72264
1946	72265–73369
1947	73370–74079
1947	74080–80555?
1948	80556–85833
1948	85834–92294
1949	92295–102392
1950	102393–193578
1951	114527–118673
1952	118674–122309
1953	122310–124200
1954	124201–124710
1955	124711–126496
1956	126497–127185
1957	127186–127461

Model IB, 1946 and later

Previous IB tractors included in B serial numbers.

1946	1001–1002
1947	1003–1009
1948	1010–1281
1949	1282–1555
1950	1556–1878
1951	1879–2118
1952	2119–2567
1953	2570–2847

Production continued until 1958.

Model C

1940	1–111

(First 9 are probably preproduction tractors, as they have some unique features.)

1941	112–12388
1942	12389–18781
1943	18782–23907
1944	23908–30694
1945	30695–36377
1946	36378–39167
1947	39168–51514
1948	51515–68280
1949	68281–80517
1950	80518–84030

Model CA

1950	14–321
1951	322–10538
1952	10539–22180
1953	22181–31423
1954	31424–32906
1955	32907–37202
1956	37203–38617
1957	38618–38976
1958	38977–39513

Model RC

1939	4–4391

(Note: 1–3 are probably preproduction tractors.)

1940	4392–5416
1941	5417–5504

Model G

1948	6–10960

Note: tractors before number six are known to exist.)

1949	10961–23179
1950	23180–24005
1951	24006–25268
1952	25269–26496
1953	26497–28035
1954	28036–29035
1955	29036–29976

(GII tractors produced in the early 1980s were probably similar only in appearance to first Gs.)

INDEX